Behaviormetrics: Quantitative Approaches to Human Behavior

Volume 18

Series Editor

Akinori Okada, Professor Emeritus, Rikkyo University,
Tokyo, Japan

This series covers in their entirety the elements of behaviormetrics, a term that encompasses all quantitative approaches of research to disclose and understand human behavior in the broadest sense. The term includes the concept, theory, model, algorithm, method, and application of quantitative approaches from theoretical or conceptual studies to empirical or practical application studies to comprehend human behavior. The Behaviormetrics series deals with a wide range of topics of data analysis and of developing new models, algorithms, and methods to analyze these data.

The characteristics featured in the series have four aspects. The first is the variety of the methods utilized in data analysis and a newly developed method that includes not only standard or general statistical methods or psychometric methods traditionally used in data analysis, but also includes cluster analysis, multidimensional scaling, machine learning, corresponding analysis, biplot, network analysis and graph theory, conjoint measurement, biclustering, visualization, and data and web mining. The second aspect is the variety of types of data including ranking, categorical, preference, functional, angle, contextual, nominal, multi-mode multi-way, contextual, continuous, discrete, high-dimensional, and sparse data. The third comprises the varied procedures by which the data are collected: by survey, experiment, sensor devices, and purchase records, and other means. The fourth aspect of the Behaviormetrics series is the diversity of fields from which the data are derived, including marketing and consumer behavior, sociology, psychology, education, archaeology, medicine, economics, political and policy science, cognitive science, public administration, pharmacy, engineering, urban planning, agriculture and forestry science, and brain science.

In essence, the purpose of this series is to describe the new horizons opening up in behaviormetrics — approaches to understanding and disclosing human behaviors both in the analyses of diverse data by a wide range of methods and in the development of new methods to analyze these data.

Keiko Nakao
Editor

Multidimensional Aspects of Occupational Segregation

Time Series and Cross-National Comparisons

Springer

Editor
Keiko Nakao
Tokyo Metropolitan University
Hachioji-City, Tokyo, Japan

ISSN 2524-4027 ISSN 2524-4035 (electronic)
Behaviormetrics: Quantitative Approaches to Human Behavior
ISBN 978-981-99-8512-8 ISBN 978-981-99-8513-5 (eBook)
https://doi.org/10.1007/978-981-99-8513-5

This Springer imprint is published by the registered company Springer Nature Singapore Pte Ltd.
The registered company address is: 152 Beach Road, #21-01/04 Gateway East, Singapore 189721, Singapore

Contents

Introduction

Keiko Nakao and Aya Wakita

1 Meritocracy in Post-industrial Societies

It had been believed that the process of industrialization would promote meritocracy in many societies. Higher technological innovations would create higher level of industries, and such industries would require jobs that need higher human capital. In highly industrialized societies, people with higher level of education and with higher skills would be needed. Therefore, it was thought that industrialization would promote meritocracy.

In the area of social stratification research, it had been taken for granted that the level of a person's education would affect his/her occupational status (e.g., Blau and Duncan 1967; Treiman 1970). The Blau–Duncan classic model for status attainment established that higher education as well as higher father's social status leads to high-status jobs. That principle has been the case in industrialized meritocratic societies. In other words, in today's performance-based society, it has been believed that a high level of education is the ticket to attaining a high occupational status. As for the social stratification literature, however, Featherman et al. (1975) proposed a hypothesis stating that industrialized countries with a nuclear family system and market economy showed similar levels of relative mobility. Later, this hypothesis was tested and supported (Erikson and Goldthorpe 1992). The idea that industrialization promotes meritocracy may have been applied during the period when the countries were going

K. Nakao (✉)
Tokyo Metropolitan University, 1-1 Minamiosawa Hachiji-Shi, Tokyo 192-0397, Japan
e-mail: nakao@tmu.ac.jp

A. Wakita
Division of Human Sciences, Ochanomizu University, 2-1-1 Otsuka, Bunkyo-ku, Tokyo 112-8610, Japan
e-mail: wakita.aya@ocha.ac.jp

K. Nakao (ed.), *Multidimensional Aspects of Occupational Segregation*,
Behaviormetrics: Quantitative Approaches to Human Behavior 18,
https://doi.org/10.1007/978-981-99-8513-5_1

through industrialization; it may not be always clear after industrialization (post-industrialization). However, given the effects of education on occupation, it would be possible that countries become more meritocratic even during post-industrial societies.

Post-industrialization is said to be characterized by more growth in occupations in the service sector, industry sophistication, and women's labor participation, which would promote meritocracy (Bell 1973; Esping-Andersen 1999). Countries with higher post-industrialization would have more jobs that require high educational levels, which will increase the importance of achieving higher education. In fact, many nations are experiencing an expansion of higher education. Furthermore, among the those with bachelor's degrees, it was reported that effects of socioeconomic origins are not as important (Torche 2011). From the perspective of educational premium, meritocracy is strong in post-industrial societies. However, if we focus on gender, could we say that meritocracy is still strong?

Expansion in service industry does not necessarily require workers with high human capital. During the age when countries were going through industrialization, lower manual workers were much in demand; however, post-industrialization would require more workers that are lower non-manual (Esping-Andersen 1999). Diminishing manufacturing industries and expanding of service industries would promote hiring routine non-manual workers such as service workers and sales workers. Female workers are in demand in order to fill these vacancies.

There has been a series of research conducted on the relationship between gender and occupation in the area of occupational gender segregation. For example, Charles and Grusky (2004) pointed out that there are two possible features in terms of the effects of gender during the era of post-industrialization, namely service-sector expansion and economic rationalization. According to Charles and Grusky, service-sector expansion changed industrial structure that would increase jobs that are perceived to be "appropriate for women." Such changes made unpaid work that women used to assume inside home for the family payable business, namely interpersonal service jobs. For this type of work, it is necessary for workers to have emotional labor or interpersonal skills that women were said to possess. Office workers and administrators in service industry have also became increasingly feminized. Thus, post-modernized nations tend to have more female non-manual workers and horizontal occupational gender segregation, indicating that women tend to be allocated to non-manual occupations and men to manual occupations.

It is not only unmarried females who were hired to assume such lower non-manual jobs. Married women also participated in a way of part-time or flexible work. Workplaces for routine non-manual workers have become female-oriented and dominated by wives and mothers with domestic responsibilities. Feminizing the lower non-manual sector strengthened vertical occupational gender segregation among non-manual workers: upper non-manual occupations are occupied by men and lower non-manual occupations by women.

On the other hand, economic rationalization would promote the division of labor and reduce self-employment in sales or small businesses, and would increase the number of large businesses. Therefore, economic rationalization has created routine, unskilled, and non-manual jobs for women. This also contributes to the increasing demand for women in non-manual occupations, and vertical occupational gender segregation in non-manual sectors.

However, service sector expansion and economic rationalization have also created managerial occupations in service industries and new opportunities for elite women. In addition to Charles and Grusky's explanation, we can emphasize that the number of professional and technical (associate professional) workers related to "feminine" skills is increasing during the post-industrialization era, especially in the medical and welfare industries. These changes may enhance meritocracy and weaken vertical occupational gender segregation in non-manual workers if highly educated women are allowed to attain upper non-manual occupations.

According to an international comparison (Charles and Grusky 2004), gender egalitarianism is associated with high levels of horizontal occupational segregation and low levels of vertical gender segregation in the non-manual sector. On the other hand, post-industrialization increases horizontal segregation and vertical segregation among non-manual occupations. Post-industrial countries have a high proportion of women among managerial, clerical, and service/sales workers and a high proportion of men among craft, operative, and labor workers.

In short, the relationships between educational attainment and occupation are still observed today; it does not necessarily mean that meritocracy is strong in post-industrial nations, once gender is considered as an attribute. Rather, post-industrialization enhances an overall horizontal gender occupational segregation and a vertical gender occupational segregation within non-manual occupations. For better-educated females, however, there would be a better chance to obtain upper non-manual occupations.

2 Analyses Focusing on Japan

In this book, therefore, we would examine the occupational segregation focusing on gender and education within Japan and compare it with other countries. What would we clarify by conducting such empirical analyses? By focusing on educational attainment and gender, we would first be able to clarify one of the important issues. That is whether an occupational gender segregation happens at all educational levels, including both higher and lower educational levels. We can examine the educational premiums for men and women and their changes caused by the industrial structure from a gender perspective.

Furthermore, by examining with other countries' data, we would be able to determine whether Japan is a unique case or the situation is common to all post-industrialized nations. We will proceed with our analyses focusing on Japan first. There are two reasons for that. First, compared with other industrialized countries, gender inequalities are quite large in Japan. As with other countries, women's work participation is increasing. However, the laws and work regulations are based on familalism, more so compared with other nations (Esping-Andersen 1999). Thus, women's status at work remains low in Japan. There is a greater percentage of women working as non-regular workers: more than half of the working women are non-regular workers such as part-time workers. There is the large wage gap between non-regular workers and regular workers. Women's careers tend to get interrupted due to such work regulations under familialism (Brinton 2001). Therefore, women are likely to obtain non-regular jobs, which leads to female marginalization at work. Furthermore, even among regular workers, the gender wage gap is pronounced (Yamaguchi 2019).

Secondly, Japan has a different situation compared to the United States and European post-industrial nations from the perspective of industrial structure. In the past, it had been observed that Japanese occupational gender segregation is characterized by more women working in manual occupations (Charles et al. 2004). Thus, horizontal occupational gender segregation is relatively small. Oda et al. (2014), that analyzed the statistical data in Japan using models from Charles and Grusky (2004), pointed out that Japan's post-industrialization is somewhat weak. Table 1 presents the employment by industry in 16 societies, consisting of 14 societies analyzed in chapter "An International Comparison of Occupational Segregation by Sex: Through a Comparison of 14 Countries" and 2 in chapter"Comparison of Occupational Segregation in East Asia: Analyzing the Structure of Segregation by Educational Background and Gender" (South Korea and Taiwan) published by ILO (2022). Even in 2019, Japan shows a lower percentage of service sector workers than countries like the U.S., the UK, the Netherlands, and France, which supposedly experienced post-industrialization. Japan is at the similar level as Korea and Italy.

Japan is characterized by strong gender inequality and a middle level of post-industrialization. Therefore, it is very meaningful to clarify the multidimensional aspects of occupational segregation focusing on Japan. The effects of (post-) industrialization might be totally different depending on educational level. Speaking more concretely, under both industrialization and post-industrialization, highly educated women would be able to get chances having higher occupational statuses. However, post-industrialization would give women with lower education lower non-manual workers' positions.

As shown in Fig. 1, highly educated women have increased dramatically in the recent years in Japan. The number of 4-year university graduates is at a level close to that of males (Ministry of Education, Culture, Sports, Science and Technology Japan 2021). Figure 1 demonstrates the percentages of students who go on to 4-year universities and 2-year colleges after graduating from high schools. Under such circumstances, there are some women with highly successful occupations. Iwama (2008) pointed out the polarization of women's status. We need to find out whether Japan is as meritocratic as the US and European countries. Hence, the similarities (or dissimilarities) in occupational status between highly educated men and women must be examined. We can clarify the effects of (post-)industrialization and gender inequalities on occupational segregation focusing on Japan first and conducting international comparison in our analyses.

Table 1 Employment by sector in 2000, 2010, and 2019

	Year 2000			Year 2010			Year 2019		
	Services (%)	Industry (%)	Agriculture (%)	Services (%)	Industry (%)	Agriculture (%)	Services (%)	Industry (%)	Agriculture (%)
Vietnam	22.3	12.4	65.3	29.6	21.7	48.7	35.3	27.4	37.2
India	24.0	16.3	59.6	26.7	21.8	51.5	32.3	25.1	42.6
China	27.5	22.5	50.0	34.6	28.7	36.7	47.2	27.4	25.3
Thailand	32.2	19.0	48.8	41.1	20.6	38.2	45.7	22.8	31.4
Philippines	47.0	15.9	37.1	51.4	15.7	32.9	58.2	19.0	22.8
Malaysia	49.5	32.2	18.4	59.1	27.6	13.3	62.8	27.0	10.3
Poland	50.3	31.1	18.7	56.6	30.3	13.1	58.7	32.2	9.2
Taiwan	53.3	38.7	8.0	57.3	37.3	5.4	58.2	36.7	5.0
Hungary	59.8	33.8	6.5	64.8	30.7	4.5	63.1	32.2	4.7
South Korea	61.3	28.1	10.6	68.4	25.0	6.6	70.4	24.5	5.1
Italy	63.0	31.8	5.2	67.6	28.6	3.8	70.3	25.8	3.9
Japan	63.5	31.4	5.1	70.4	25.6	4.1	72.4	24.2	3.4
France	69.6	26.3	4.1	74.8	22.3	2.9	77.1	20.4	2.5
UK	73.3	25.2	1.5	79.6	19.2	1.2	80.9	18.1	1.0
U.S	73.9	24.4	1.6	78.9	19.6	1.4	78.8	19.9	1.4
Netherlands	75.1	21.6	3.3	79.2	17.7	3.1	81.8	16.1	2.1

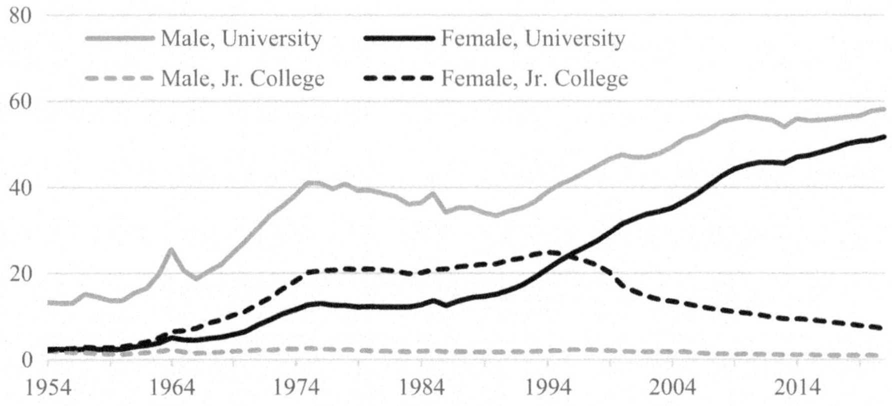

Fig. 1 Percentage of students who go on to universities or junior colleges

2.1 Methods (Correspondence Analysis)

In analyzing the data, we used correspondence analysis, focusing on the multidimensional aspects of occupations. If we used an index such as the Occupational Prestige Scores or the Socio-economic Index (SEI), we would only be able to see a unidimensional aspect of occupations, assuming that there were linear relationships between educational achievements and occupational status. We would need to overcome such assumptions. Researches in occupational segregation have shown both horizontal and vertical segregations. In addition, occupational images consist of two different dimensions: prestige and gender type (Glick et al. 1995). Occupational status is not unidimensional, but multidimensional. That is why we need to treat occupational status as categories and to use categories as small as possible.

There has been a trend in the area of social stratification research, which is called micro-class analysis. It is a response to criticisms about the current measures of social status not being appropriate because they no longer reflect the reality of the current situations (e.g., Weeden and Grusky 2012). A Japanese sociologist, Miwa (2009), points out the need to apply a more micro-level approach to clarify social disparities in Japan. That includes those analyses using more detailed occupational categories.

As Weeden and Grusky (2012) argues, it is difficult to believe that traditional broad occupational categories accurately measure contemporary hierarchical structures. Grusky, relying on Durkheim's perspective in his "division of labor theory," argued for the need to focus on disaggregated occupational categories to understand contemporary society in terms of occupations.

Class analysis based on micro occupational categories has been continuously conducted since the 2000s. It is not clear whether Grusky's assertion that Marxist class theory based on a macro perspective is no longer persuasive in contemporary society, where meritocracy is progressing, is valid. However, there is no doubt that the focus on disaggregated occupational categories has appealed to many researchers conducting class analyses to understand contemporary society.

This study is not a criticism for such macro-analysis. Instead, it empirically examines the components of occupations on which macro class analysis was based. In other words, we focus on the multidimensionality of occupations.

The premise of Grusky's argument that traditional occupational categories are becoming inadequate to capture current classes can be interpreted as the fact that the social significance of occupations in contemporary society has become increasingly diverse. This implies the need not only to subdivide occupations but also to reconfigure them. Indeed, segmentation of occupational categories is practiced in micro-class analysis. However, in light of the transformation of the social structure, including post-industrialization, it is also important to reconstruct occupational categories according to the dimensions constructed in the society in question, rather than assuming only the hierarchical structure of occupations. Especially for cross-national data analyses, we may use somewhat larger categories based on available data.

Next, we will discuss the analytical technique; correspondence analysis. This study will examine the relationship between attributes and occupations in contemporary society by disaggregating and reconstructing occupations through correspondence analysis.

What is correspondence analysis? Correspondence analysis is a multivariate statistical technique that allows us to examine the relationships between two or more variables. It estimates distance between categories as similarity or dissimilarity from the perspective of correspondence with other variables, and display them on one or more dimensional spaces. Thus, a distance between the categories indicates how similar they are.

One important merit is that it does not assume linearity in the associations. Let's look at the relationship between years in education and occupational status. When both variables are interval measures, you would normally apply regression models to analyze the data. However, correspondence analysis takes the stance that each variable is categorically measured and finds the configuration that would maximize the relationship. Other than expected dimensions such as status, education, and gender, we can examine whether there is something else that would make difference determining configurations of occupations. In other words, you could elicit how occupations differ based on the categories in education or gender.

For example, take a look at the pattern seen in Fig. 2. This is an example of taking education as the first dimension and gender as the second dimension.

Figure 2a presents a pattern in which status is approximately equal between men and women with the same educational levels. Figure 2b, on the other hand, describes a situation in which status varies by gender. Males have higher status than females. Females with high education would be the same as males with middle education. Figure 2c shows the pattern in which education affects differently for males and females. The distance between the categories is longer for males. This implies that the effect of education is greater for males than females. Finally, Fig. 2d is the pattern in which gender effect depends on educational levels. If one attains high education, gender difference is small. The theory of industrialization would have predicted this pattern, meritocracy.

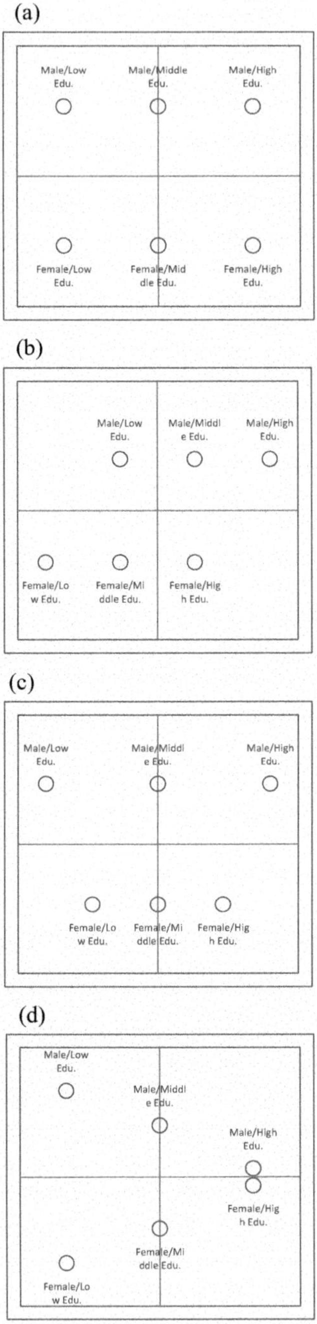

Fig. 2 Examples of configurations

Correspondence analysis is powerful when focusing on several different dimensions that characterize an occupation, as described above. Theoretically, after going through the industrialization and the post-modernization within each country, it is not surprising that the meaning of occupations has become more complicated and varied. Other than status, we need to investigate what kind of dimensions occupations have.

By conducting this type of analyses, we will be able to obtain results on multidimensional occupational segregation associated with gender and educational levels. Dimensions obtained from our analyses may not necessarily correspond with horizontal/vertical segregation. However, these analyses would certainly provide important knowledge for research on occupational gender segregation.

2.2 Brief Description of Chapters

Here, we will briefly introduce the chapters of this book. Chapters "Multidimensional Structure of Occupational Segregation by Education and Gender: Time Series Comparison in Japan from 1980 to 2010" through "Occupation Segregation for Immigrants in Japan: Using Custom-Made Census Data in 2010" focus on analyses in Japan and demonstrate the Japanese features. In chapter "Multidimensional Structure of Occupational Segregation by Education and Gender: Time Series Comparison in Japan from 1980 to 2010", we first investigate a time-series comparison in Japan from 1980 to 2010. We then identify what has been happening in Japan. In chapter " Occupational Segregation by Education and Gender in Japan: Focusing on Employment Status and Marital Status", we looked at the effect of gender focusing on employment status and marital status. They are important factors related to women's labor participation. We also examined immigrants in Japan, which has recently increased in number (chapter "Occupation Segregation for Immigrants in Japan: Using Custom–Made Census Data in 2010"). Analyses in chapter "Occupation Segregation for Immigrants in Japan: Using Custom-Made Census Data in 2010" include not only gender but also ethnicity, pointing the importance that there are various attributes to consider in Japan.

The following chapters focused on the cross-cultural analyses between Japan and other countries: the United States (chapter "Multidimensional Structure of Occupational Segregation in Japan and the United States: Comparison of Trends from 1980 to 2010"), East Asian societies (chapter "Comparison of Occupational Segregation in East Asia: Analyzing the Structure of Segregation by Educational Background and Gender"), and 14 countries, including European countries, Asian countries, and the U.S. (chapter "An International Comparison of Occupational Segregation by Sex: Through a Comparison of 14 Countries"). Detailed comparisons between the US and Japan reveal the effects of post-industrialization and gender egalitarianism (chapter "Multidimensional Structure of Occupational Segregation in Japan and the United States: Comparison of Trends from 1980 to 2010"). While the US is more advanced in post-industrialization and gender egalitarianism, Japan is not progressing much. The chapter clearly presents the difference. East Asian societies (chapter "Comparison of Occupational Segregation in East Asia: Analyzing the Structure of Segregation by Educational Background and Gender") have experienced so-called "compressed modernity." The difference among them would suggest why such difference would be observed despite their similarities. And finally, comparisons among 14 countries with varying degrees of industrialization/post-industrialization would lead to the conclusion as to how industrialization/post-industrialization will make occupational segregation by education and gender in each country unique (chapter "An International Comparison of Occupational Segregation by Sex: Through a Comparison of 14 Countries").

With this manuscript, overall, the multidimensional occupational segregation will become clearer by analyses in Japan and international comparisons. This will help us to place Japan within the entire world.

References

Bell, Daniel. 1973. The coming of post-industrial society. In *Social stratification*, 4th ed., ed. David B. Grusky, and Katherine R. Weisshaar, 2014, 1066–1077. Boulder: Westview Press.

Blau, Peter M., and Otis Dudley Duncan, with the collaboration of Andrea Tyree. 1967. The process of stratification. In *Social stratification*, 4th ed., ed. David B. Grusky, Katherine R. Weisshaar 2014, 486–497. Boulder: Westview Press.

Brinton, Mary C. 2001. Married women's labor in East Asian economies. In *Women's working lives in East Asia*, ed. Mary C. Brinton, 1–37. Stanford: Stanford University Press.

Charles, Maria, and David B. Grusky. 2004. *Occupational ghettos*. Stanford: Stanford University Press.

Charles, Maria, Mariko Chang, and Joon Han. 2004. Gender and age in the Japanese labor market, 1950–1995. In *Occupational ghettos*, ed. Maria Charles, and David B. Grusky, 179–212. Stanford: Stanford University Press.

Erikson, Robert, and John H. Goldthorpe. 1992. *The constant flux: A study of class mobility in industrial societies*. Oxford: Oxford Univ Press.

Esping-Andersen, Gøsta. 1999. *Social foundations of postindustrial economies*. Oxford: Oxford University Press.

Featherman, David L., Lancaster Jones, and Robert M. Hauser. 1975. Assumptions of social mobility research in the US: The case of occupational status. *Social Science Research* 4: 329–360.

Glick, Peter, Korin Wilk, and Michele Perreault. 1995. Images of occupations: Components of gender and status in occupational stereotypes. *Sex Roles* 32: 565–582.

ILO. 2022. "ILOSTAT". https://ilostat.ilo.org/data/.

Iwama, Akiko. 2008. *Josei no Shugyo to Kazoku no Yukue* (Women's Work and Future of Family), Tokyo: University of Tokyo Press.

Ministry of Education, Culture, Sports, Science and Technology Japan. 2021. "Gakko Kihon Chosa (School Basic Survey)". https://www.mext.go.jp/b_menu/toukei/chousa01/kihon/1267995.htm.

Miwa, Satoshi. 2009. Shakaiteki Kakusa wo Toraeru tameno Micro-Class Approach (A new class mapping using micro-class Shema from JGSS cumulative dataset). *JGSS Research Series* 8: 1–12.

Oda, Akeko, Reiko Yamato, and Hiroshi Taroumaru. 2014. Seibetsu-shokuiki-bunri ha chiiki niyotte dou kotonaruka (An inter regional comparison of occupational gender segregation in Japan). In *Higashi-Asia no roudou-shijo to shakai-kaisou* (*Labor market and social stratification in East Asia*), ed. Hiroshi Taroumaru, 137–168. Kyoto: Kyoto University Press. (in Japanese).

Torche, Florencia. 2011. Is a College Degree Still the Great Equalizer? Intergenerational Mobility across Levels of Schooling in the United States. *American Journal of Sociology* 117(3): 763–807.

Treiman, Donald J. 1970. Industrialization and social stratification. *Sociological Inquiry* 40 (2): 207–234.

Weeden, Kim A., and David B. Grusky. 2012. The three worlds of inequality. *American Journal of Sociology* 117 (6): 1723–1785.

Yamaguchi, Kazuo. 2019. *Gender Inequalities in the Japanese Workplace and Employment: Theories and Empirical Evidence*, Singapore: Springer.

Multidimensional Structure of Occupational Segregation by Education and Gender: Time Series Comparison in Japan from 1980 to 2010

Takuya Hayashi

1 Introduction

Occupation is essentially a nominal division according to the type of work with which people engage. Occupational segregation indicates the degree and pattern of the cohesion and division among various occupations and is crucial to understanding the structure of occupational inequality. It is necessary to investigate and analyze distances among occupations, based on some set criteria. For example, "occupational prestige," a unidimensional status indicator, is a measure given to each occupation based on the objective evaluation of people in a certain society (e.g., Nakao and Treas 1994; Tsuzuki 1998; Genji 2018). It has also been deployed from several different perspectives to capture association/segregation structures based on the distance between occupations. Focusing on cognition likewise, Hayashi (2017) adopted another criterion of the subjective cognition of people who engage in a certain occupation and evaluate other occupations by a sense of affinity/distance. Furthermore, there have also been other methods based on occupational association with social relations such as friends and spouses (Laumann and Guttman 1966; Prandy 1990; Bakker 1993; Prandy and Lambert 2003; Chan and Goldthorpe 2004), based on the amount of intra-/inter-generational mobility (Rytina 1992; Okada and Imaizumi 1997; Kondo 2006; Toubøl and Larsen 2017).

These studies, using criteria such as strength of association (cognitive affinity, social relationship, amount of mobility), reflect the viewpoint of *deriving from the analysis of the structure of occupational segregation inherent therein.* On the other hand, the index of occupational status such as prestige score that is used to analyze the linear relationship with educational status is a scale with an external criterion, in

T. Hayashi (✉)
Division of Humanities and Social Sciences, Nara Women's University, Kitauoyanishi-Machi, Nara City, Nara 630-8506, Japan
e-mail: t-hayashi@cc.nara-wu.ac.jp

K. Nakao (ed.), *Multidimensional Aspects of Occupational Segregation*, Behaviormetrics: Quantitative Approaches to Human Behavior 18,
https://doi.org/10.1007/978-981-99-8513-5_2

that it has not been derived from the analysis, but *given with the scale value before the analysis*.[1] However, when exploring the relationship between occupation and education, it is also possible to derive *from the analysis* the structure of occupational segregation inherent in homogeneity/heterogeneity in terms of educational status from the former viewpoints. By doing so, it is expected to reveal the kind of linkage that exists between education and occupation besides their linear relationship using the unidimensional scale of occupational status (e.g., prestige score) as described above.

As for occupational segregation by educational status, it is necessary to take into account gender differences. The phenomenon of the difference in occupational composition between men and women is known as "occupational gender segregation" and has been extracted as an independent dimension in empirical studies on multidimensional occupational segregation (Chan and Goldthorpe 2004; Hayashi 2017). Such gender (sex) segregation is regarded as an issue open to verify whether occupational segregation by education is the same between males and females or, in other words, whether the segregation occurs as a result of a combination of sex and education.

It is also important from a theoretical perspective to focus on both. As Treiman (1970) once argued, it has been believed that, in an industrialized society, the influence of ascriptive status or attributes such as socio-economic status of one's origin (i.e., one's parental status) on occupational attainment would diminish, and that of achieved status such as education or qualifications would predominate. Note that among ascriptive attributes gender differences in occupational attainment is controversial. When a unidimensional scale of occupational status such as prestige is used, the difference in average status is not so large. However, occupational segregation occurs in a fairly clear way using occupational categories as described above.[2] In light of this, the differences in occupational attainment by gender (as an ascriptive status) need to be separately examined from that by education (as an acquired status), handling carefully the index of occupational status.

For examination of the above, it is desirable to obtain data on occupation as finely divided as possible. Considering that institutions such as skill formation, recruitment, and licensing are arranged at a smaller occupational group level as argued by microclass advocates (Grusky and Sørensen 2001; Grusky 2005; Miwa 2009), there may be educational and/or gender variations within the existing occupational classification (big-class). Additionally, when treating the subdivided occupational classification, it seems effective to extract and interpret a few principal axes briefly representing the segregation structure and explore changes in time based on those axes. For this purpose, I applied correspondence analysis to a contingency table and expressed the relationship among occupational categories on the dimensional space.

2 Data and Method

2.1 *Data and Variables*

The data to be analyzed in this research was obtained from the National Population Census in Japan conducted by the Ministry of Internal Affairs and Communications. Although the surveys are carried out every 5 years, information about subjects' education has been obtained every 10 years. I use the corresponding data of the last census in 2010 and the previous three censuses in 2000, 1990, and 1980. Note that information on the minor occupational classification, which will be described in detail below, was obtained from the sampled data "Detailed Sample Tabulation" extracted from the census (in 2010, the sampling ratio was 9.8%). In spite of a smaller number of subjects than the original census, because the sample size is much larger than other sample surveys, the estimation could be stable for the worker population in over 200 minor occupational groups.

In addition, as described later, this research requires the number of workers to be classified by sex, age group, and education at each year. However, I cannot get all of that information from the report published or from the Internet. Therefore, I got the National Statistics Center to provide the required aggregate tables by utilizing the "order-made summarization" service.[3]

In the census data used in this research, the minor classification of occupation has been partially changed by the survey year; the number of occupational titles is 292 for the 1980 data, 293 for 1990, 292 for 2000, and 231 for 2010 (excluding "not classifiable"). It would be valid to use the classification of each year as is if I analyze the data of a single year. However, this research focuses on the changes in time and needs a common classification to be applied across the four annual data sets. Therefore, my colleagues and I set 157 categories that integrated these categories as the basis of occupational classification for analysis. The combination of sex and education is used in a total of eight categories comprising 2 (male/female) × 4 (junior high school/senior high school/junior college/university [including graduate school]) categories.[4] Furthermore, by dividing the core workers (20–59 years old at each year) into four age groups by 10 years, I prepared a contingency table of 157 categories of occupation × 8 categories of personal attributes (i.e., sex and education) respectively as a dataset to be analyzed.

2.2 *Analytical Framework*

As mentioned above, in this study the correspondence analysis was applied to the contingency table. Correspondence analysis is a method of reproducing the distance between categories of two (or more) variables as positions on one or more axes, estimating coordinate points of the categories (Benzécri 1992; Greenacre 1993; Clausen 1998). The dimension is composed by the number of axes, therefore, combining two or more axes leads to a focus on multidimensional space. The farness/nearness on an axis or space represents the dissimilarity/similarity between the categories. In terms of the variables I deal with in this chapter, if the occupational composition of a given education category A and B is close, these categories will be placed close together in the axis or space, and if their occupational compositions are very different, they will be placed far apart. The placement of the categories is called configuration and the position of each category on the axes is estimated as quantified coordinate point. Using the variables of occupation and attributes (sex-education), the coordinate points of the former categories in the extracted axes or space will reveal the segregation structure of occupations based on the traits of gender and educational status; occupations that are closely/separated in the axes or space indicate that they comprise people with the same/different sex and education.

Correspondence analysis could be applied to the contingency table composed of occupation × personal attributes for each age group in a year, which would derive a different dimensional structure for each. If there is a large fluctuation in the structure, it might be useful to compare the results obtained from that method. However, since its structure changes only gradually, as Hayashi et al. (2017) confirmed,[5] it is more useful to generate a common dimensional structure throughout all time points and age groups (hereafter, "year-age group(s)") and to explore the change that has been occurring within the structure. Therefore, in this research, by deriving the common dimensional space as a whole, I will examine changes in occupational configuration in that space.

In exploring the above, two methods seemed conceivable. The first is a method to fix the space with the *specific* year-age group and then set the configuration of other groups as supplementary points in the space; the second is a method to derive the space with *all the year-age groups pooled* and to confirm each of their coordinates. In the former method, there is a strong premise that the common dimensional space is determined by that of the *specific* year-age group selected. Thus, the results might differ depending on which year-age group is set as the base point. To avoid this ambiguity, I adopted the latter method in this study.

The image to pool the data is shown in Table 1. By setting the year-age group variable as T, the occupation variable as A, and the sex-education variable as B, a stacked contingency table $(AT) \times B$ was created (Bijleveld and van der Burg 1998: 85–89; Kimiyama 2005: 30–33). Note that t is a category of T, i is a category of variable A, j is a category of variable B, f is a cell frequency, and N is a marginal frequency or grand total. When applying correspondence analysis to this data, the configuration of the categories of variable B is common throughout year-age groups t, whereas the configuration of the categories of variable A is different depending on t (Bijleveld and van der Burg 1998: 88). Therefore, the results of analysis are expected to reveal how the position of each occupation varies (changes) by year-age groups in the dimensional space in which the categories of sex-education are fixed.

Table 1 Image of the stacked contingency table from the pooled data

T	*A*	*B*				*N*
		1	2	..	*j*	
1	1	f_{111}	f_{112}	..	f_{11j}	$N_{11\cdot}$
	2	f_{211}	f_{212}	..	f_{21j}	$N_{21\cdot}$
	:	:	:		:	:
	i	f_{i11}	f_{i12}	..	f_{i1j}	$N_{i1\cdot}$
2	1	f_{121}	f_{122}	..	f_{12j}	$N_{12\cdot}$
	2	f_{221}	f_{222}	..	f_{22j}	$N_{22\cdot}$
	:	:	:		:	:
	i	f_{i21}	f_{i22}	..	f_{i2j}	$N_{i2\cdot}$
	:	:	:	..	:	:
t	1	f_{1t1}	f_{1t2}	..	f_{1tj}	$N_{1t\cdot}$
	2	f_{2t1}	f_{2t2}	..	f_{2tj}	$N_{2t\cdot}$
	:	:	:		:	:
	i	f_{it1}	f_{it2}	..	f_{itj}	$N_{it\cdot}$
N		$N_{..1}$	$N_{..2}$		$N_{..j}$	$N_{...}$

3 Dimensional Space from Correspondence Analysis

3.1 Dimensions Derived and the Configuration of Attributes

As a result of applying correspondence analysis to the contingency table described above, eigenvalues and cumulative contribution ratios of each number of dimensional solutions were obtained as shown in Fig. 1.[6] The eigenvalue of a dimension represents its relative importance in the relation of variables analyzed (i.e., occupation and attributes). The maximum number of dimensions and eigenvalues is the smaller of the number of rows or columns in the contingency table minus one. In this analysis, the smaller is eight categories of education-gender combination and seven dimensions are extracted. The sum of eigenvalues is called "total inertia," which represents the magnitude of association between variables in the contingency table and equivalents to Pearson's chi-square statistic divided by the total number of individuals. The ratio of eigenvalue of a dimension to total inertia corresponds to the contribution ratio. Checking the scree of eigenvalues, the "elbow" that turns from a sharp fall to a gradual decrease could be seen at the third dimension, indicating explanatory power of three or more dimensions does not change very much. Therefore, I adopt a two-dimensional solution (the cumulative contribution is 61.6%).

The configuration of sex-education categories in the two-dimensional space is shown in Fig. 2a. In the first dimension (horizontal axis), "female-university" is located in the positive pole, and "male-junior high school" is located in the negative pole. However, in the entire two-dimensional space, it is clear that the contrast

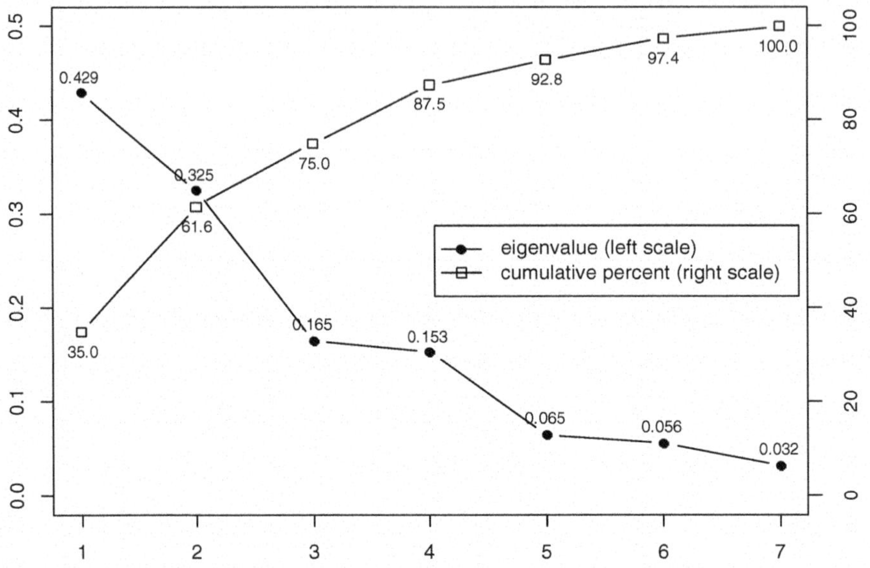

Fig. 1 Eigenvalues and contributions in correspondence analysis

between the third and the first quadrants is segregation by educational status, and that of the second and the fourth quadrants is segregation by gender. Therefore, Procrustes rotation[7] is carried out (Okada and Imaizumi 1994: 134–141; Gower and Dijksterhuis 2004) in order to facilitate the interpretation of each dimensional axis. The target matrix is set in the first axis (horizontal axis) as −1 for junior high school, −0.5 for senior high school, 0.5 for junior college, and 1 for university so that the axis represents educational segregation; it is set in the second axis (vertical axis) as 1 for males and −1 for females so that the axis represents gender segregation. The configuration after rotation is as shown in Fig. 2b. As can be seen from the figure, the segregation by both variables is almost independent; occupational segregation by educational status for males is not equivalent to that for females. Thus, when focusing on one segregation it will be necessary to retain attention on the other segregation.

3.2 *Structure of the Configuration of Occupations*

Next, I looked at the configuration of occupations in the two-dimensional space. The configuration does not comprise original coordinate points of occupations obtained by correspondence analysis, but those obtained after that by rotating the axes to represent educational and gender segregation as described above. Therefore, occupations dominated by highly educated/poorly educated people will be located in the positive/negative direction on the horizontal axis (dimension 1) and those dominated by men/

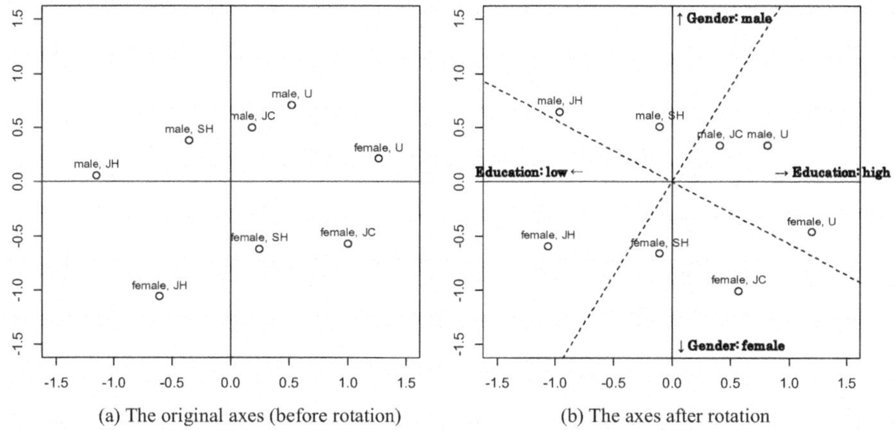

Fig. 2 Configuration of sex-education categories in a two-dimensional space. JH: junior high school, SH: senior high school, JC: junior college, U: university. The dotted lines in **b** indicate the axes before rotation

women will be located in the positive/negative direction on the vertical axis (dimension 2). In interpreting the configuration of occupations, hereinafter, the horizontal and vertical axes are called the "educational qualification" and "gender" dimensions, respectively.

From the analysis, 2,512 coordinate points comprising 157 occupations × 4 years × 4 age groups were obtained, and I examined each year-age group individually instead of as a whole. Figure 3 shows the centroid of each year and age group in the two-dimensional space. The centroid is the mean of coordinate points weighted by population included in the concerned group (e.g., 20s in the year of 1980). The centroids of years shift from the second to the fourth quadrant as it becomes more recent, which indicates that the working population has shifted toward highly educated females in recent years, considering the configuration of attributes seen before. Analyzing the centroids of age groups each year reveals that younger workers are more highly educated (but the difference shrinks in 2010) and that there are more men in their 30s than in other age groups. The latter is thought to reflect the tendency for women of that age to temporarily leave their jobs because of marriage, childbirth, or child-rearing.

The total number of year-age groups is 16 (year 4 × age group 4), and Fig. 4a–d shows the configuration of 157 occupations in the four year-age groups selected from among them. Note that each occupation is represented by symbols "A" to "K" based on the major groups of the Japan Standard Occupational Classification (JSOC) to which each occupation belongs.

While the dispersion of occupations differs between the four groups shown here, the relative position of the major groups of the JSOC is generally common. Many of the professional and technical occupations (A) are located in the right direction of the space, which shows that they are composed of highly educated workers. Typical

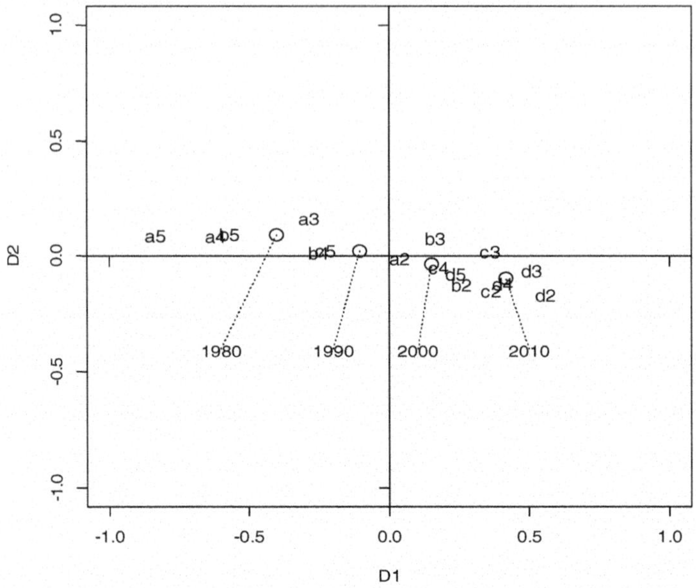

Fig. 3 Centroids of years and age groups. ○: Centroid of each year. Alphabet-number combinations represent age groups at each year. [Year] a: 1980, b: 1990, c: 2000, d: 2010. [Age] 2: 20–29, 3: 30–39, 4: 40–49, 5: 50–59

occupations are scientists, physicians, pharmacists, judges, prosecutors, lawyers, and teachers. In addition, some of the professional and technical occupations, specifically nurses, midwives, public health nurses, childminders, and kindergarten teachers are located in the lower direction of the space, which shows that they are composed of females and mainly junior college graduates, not 4-year university graduates. Managerial occupations (B) are located in the upper right direction of the space, and these consist of males and highly educated workers. Many of the clerical occupations (C) are located at the center of the space, indicating that they do not consist of specific attributes. However, some located toward the bottom consist mainly of female workers. Sales occupations (D) are located at the center on the horizontal axis, but are dispersed along the vertical axis, confirming that the occupation is divided into those with many males and those with many females. Service occupations (E) (e.g., customer service, hairdresser) are generally located in the lower left side of the space and comprise female, poorly educated workers. Protective occupations (F) are located in the upper center of the space, consisting of male workers. Many agricultural occupations (G) are located in the upper left side of the space and mainly comprise male, poorly educated workers. Several occupations related to the manufacturing process (H), transportation and machine operation (I), and construction (J) are also located in the upper left side of the space and also mainly comprise male, poorly educated workers. Note that some occupations included in the manufacturing process are located at the center on the vertical axis and are composed not solely of male

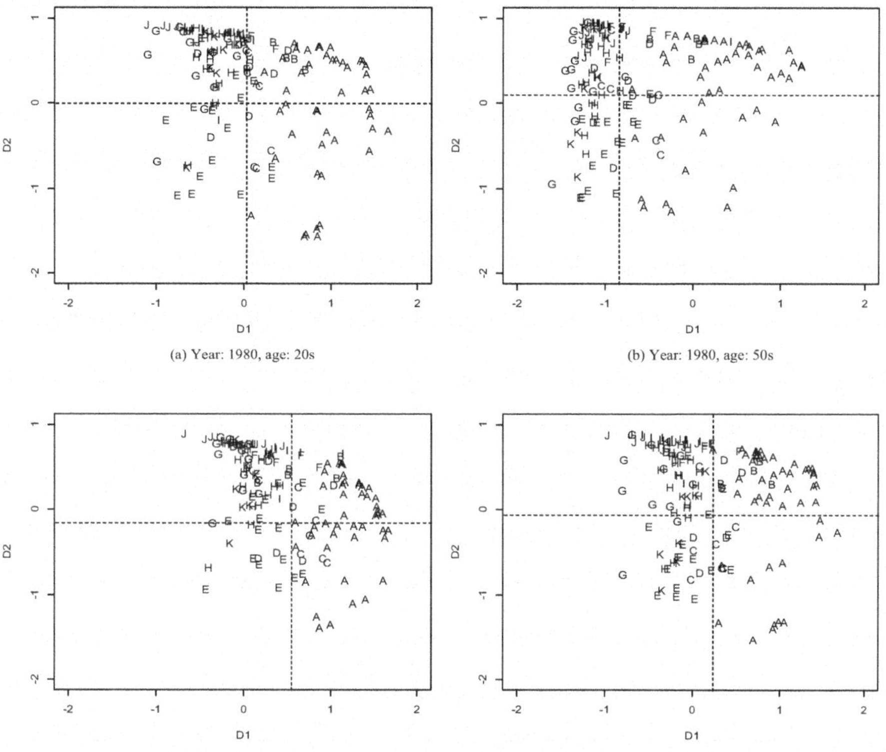

(a) Year: 1980, age: 20s

(b) Year: 1980, age: 50s

(c) Year: 2010, age: 20s

(d) Year: 2010, age: 50s

Fig. 4 Configuration of occupations in a two-dimensional space. The dotted lines represent the centroid of each axis of the year-age group

workers. Occupations involving carrying, cleaning, packaging, and related work (K) are located on the left side of the space and are composed of poorly educated workers, among which there are those with more male workers (upper left direction) and those with more female workers (lower left direction).

Next, I examined how the arrangement of occupations, in particular on the horizontal axis representing educational qualification, was related to the status scale on occupations. Table 2 shows the correlation coefficient in each year-age group between the coordinate points of occupations in the educational qualification dimension (D1) and the occupational prestige scores. They are around 0.7 and in 1980 it showed a higher correlation among the elder than among the younger age groups, whereas this tendency disappeared in 2010. From these results, it could be said that occupational segregation by education corresponds to a certain extent to the order of occupational prestige and that this correspondence is not as strong as before in the 40s and 50s.

Table 2 Correlation coefficients between the coordinate points in the educational qualification dimension (D1) and the prestige scores of occupations

Year	Age group			
	20s	30s	40s	50s
1980	0.718	0.758	0.781	0.799
1990	0.732	0.706	0.742	0.764
2000	0.709	0.718	0.692	0.729
2010	0.701	0.717	0.721	0.689

4 Examination of Time Change

4.1 Changes in Configuration and Contribution

With regard to the structure of occupational segregation as outlined in the previous section, I examined the changes over time from the following two viewpoints: how the configuration as shown in Fig. 4 has changed and how the importance of each dimension (educational qualification/gender) has changed.

For the first viewpoint, the coordinate points of each dimension with the units of 157 occupations were examined by revealing the correlation between adjacent years in the same age group (Table 3). It was confirmed that all correlations were extremely high in both dimensions and that the basic structure (the relative position in the dimensional space) has been maintained through the times.

For the second viewpoint, a comparison was made between years using "inertia," an index representing the degree of contribution of each dimension. The total inertia described earlier is decomposed into inertia of each dimension and category (note that the eigenvalue of dimension equivalents to the inertia of the dimension). The inertia of category i in the rth dimension is obtained from the product of the square of the coordinate point x_{ri} and the mass (proportion of the category) m_i, that is, from

Table 3 Correlation coefficients of coordinate points of occupations between adjacent years

		Between adjacent time points		
		1980–1990	1990–2000	2000–2010
D1	20s	0.977	0.987	0.973
	30s	0.982	0.987	0.988
	40s	0.985	0.986	0.986
	50s	0.978	0.988	0.986
D2	20s	0.973	0.982	0.979
	30s	0.982	0.983	0.984
	40s	0.977	0.987	0.986
	50s	0.968	0.986	0.988

the following Eq. (1) (Greenacre 1993: 89):

$$\text{inertia}_{ri} = x_{ri}^2 m_i = x_{ri}^2 \left(N_{it.} / N_{...}\right) \tag{1}$$

Then, the inertia representing the contribution of the whole r dimension can be obtained from this sum, that is, from the following Eq. (2):

$$\text{inertia}_r = \sum\nolimits_i x_{ri}^2 m_i = \sum\nolimits_i x_{ri}^2 \left(N_{it.} / N_{...}\right) \tag{2}$$

In this study, it is necessary to note that x_{ri} not only expresses the position of an occupation but also contains that of the year-age group. Specifically, as seen in Fig. 3, the centroids of the year-age groups are significantly different, particularly in dimension 1. When the Eq. (1) is applied to such data, the inertia of occupations included in the year-age group close to the origin (center of the space) is low, and those included in the year-age group far from the origin are high. However, focusing on the *internal variance* of each year-age group, it would be reasonable to treat the centroid of each group as its origin. Therefore, the coordinate point of an occupation is converted into the distance from the centroid of the year-age group in which the occupation is included. When the centroid of a year-age group t in the rth dimension, weighted by the respective population of the occupation, is as follows:

$$\overline{x}_{rt} = \frac{\sum_i x_{rit} N_{it.}}{N_{.t.}}$$

the distance of occupation i from the centroid is as follows:

$$x_{rit} - \overline{x}_{rt} = x_{rit} - \frac{\sum_i x_{rit} N_{it.}}{N_{.t.}}$$

By calculating inertia afresh based on the origin of each year-age group, the coordinate points in the previous Eqs. (1) and (2) are replaced with the distance from each centroid, and mass as proportion occupying the whole is replaced with mass as proportion occupying each group. Equations (1) and (2) could be rewritten as (3) and (4), respectively, as follows:

$$\text{inertia}_{ri\ \text{within\_}t} = \left(x_{rit} - \overline{x}_{rt}\right)^2 m_{i\ \text{within\_}t} = \left(x_{rit} - \overline{x}_{rt}\right)^2 \left(N_{it.}/N_{.t.}\right) \tag{3}$$

$$\text{inertia}_{r\ \text{within\_}t} = \sum\nolimits_i \left(x_{rit} - \overline{x}_{rt}\right)^2 m_{i\ \text{within\_}t} = \sum\nolimits_i \left(x_{rit} - \overline{x}_{rt}\right)^2 \left(N_{it.}/N_{.t.}\right) \tag{4}$$

This is approximately equivalent to the idea of "within-group inertia" by Greenacre (1984: 204–205), except for the mass m. When decomposing inertia as a whole into between-group and within-group inertia, the mass of each category is set as the proportion occupying the whole ($m_i = N_{it.} / N_{...}$). If the data differ in the proportion for each group, within-group inertia of a group having a large (small)

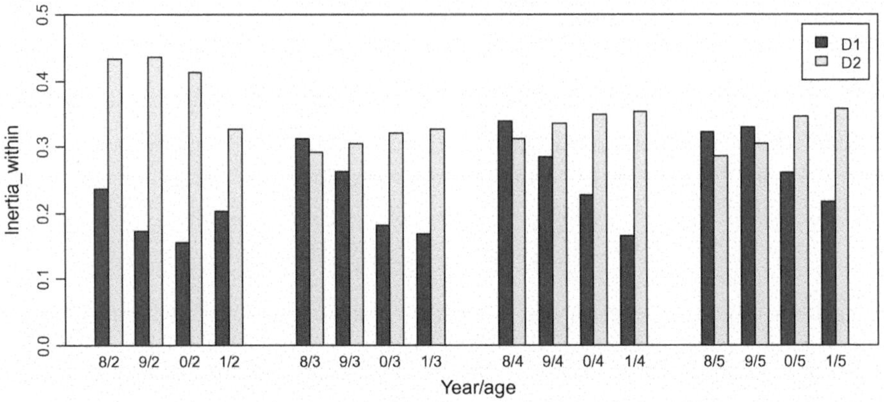

Fig. 5 Within-group inertia by year-age group. For "Year/age", for example, "8/2" represents "1980/20s"

proportion tends to be large (small). Since I want to compare the relative importance of dimensions in each year-age group in this study, however, the sum of mass of each group is set uniformly to "1" in order to adjust the difference in the population size of the group.[8] In view of this, the contribution obtained from Eqs. (3) and (4) could be called "within-group inertia adjusted for masses of groups." While it is simplified and described as "within-group inertia" hereafter, it should be noted that the value indicates the one adjusted for the difference in population size among the groups.

Figure 5 shows the within-group inertia obtained from Eq. (4) for each year-age group. According to this figure, the characteristics of and changes in each age group could be understood as follows. In the 20s, segregation by gender (D2) is greater, especially remarkably until 2000, than that by educational qualification (D1). However, as of 2010, the contribution of D1 has risen and that of D2 declined greatly; thus, the difference between them has become smaller. In the 30s–50s, similar characteristics are seen. That is, while in 1980 the contribution of D1 and D2 were comparable, that of D1 has fallen and that of D2 has risen steadily along with the passage of time. As a result, gender segregation is predominant as of 2010. It is inferred that the decrease in the contribution of D1 is a cohort effect rather than a period or aging effect. Tracing the same cohort at each year (e.g., 20s in 1980 and 30s in 1990), it was confirmed that the contribution has not changed significantly. Therefore, the contribution of D1 in the 30s–50s as of 1980 indicates the extensive segregation by educational qualification from the first job in those cohorts.

4.2 *Examination by Categorizing Occupations*

It is not easy to grasp the overall tendency by examining the kind of changes in occupational segregation by looking at 157 occupations. Thus I consolidate occupations into a few categories and focus on the centroids and contributions in a dimensional space. Of the classification of 11 categories shown in Fig. 4, occupations included in "professional and technical" (A) are widely distributed in the two-dimensional space and their centroid cannot successfully represent those located in the positive direction of dimension 1 or negative direction of dimension 2. Taking this into consideration, I divided "professional and technical" occupations into three categories (see [Appendix] for the validity) as follows:

[A1] "professional with advanced qualification," requiring national qualifications on the premise of a bachelor's degree or higher qualification, such as school teachers, researchers, legal workers, physicians, certified public accountants, and tax accountants.

[A2] "professional and technical with personal service," engaging in interpersonal and care work based on professional qualifications at medical, welfare, and child education job sites, such as nurses, public health nurses, childminders, social welfare professionals, and kindergarten teachers.[9]

[A3] "other professional and technical," not requiring professional qualifications through formal education, such as engineers, persons of religion, journalists, personal tutors, and artists.

(1) Changes in the dimension of educational qualification

Using a total of 13 categories that include the 3 abovementioned categories, Fig. 6 shows the change of the centroid of each category in the educational qualification dimension (D1). A1 (professional with advanced qualification) has been projected in the positive direction and has been substantially constant except for 50s in 1980,[10] indicating that it has kept on associating with highly educated people at almost any time. In other occupational categories, the tendency toward the positive direction could be seen, as it has become in recent years. From the corresponding configuration of educational status (Fig. 2b), for example, it is interpreted that those engaging in manual occupations (E, G–K) located in the negative direction have been shifting from junior high school graduates to senior high school graduates.

In Fig. 7, the values calculated from the following Eq. (5) are shown to further decompose "within-group inertia" into that of each occupational category:

$$\text{inertia}_{r\,\text{within\_}kt} = \sum_{i=1}^{K} (x_{rit} - \overline{x}_{rt})^2\, m_{i\,\text{within\_}t} = \sum_{i=1}^{K} (x_{rit} - \overline{x}_{rt})^2\, (N_{it.}/N_{.t.}) \quad (5)$$

What is different from the previous Eq. (4) is the summing up of occupations i (1, 2, …, K) included in a category k.

While it is common for the contribution of A1 (professional with advanced qualifications) to be large in any year-age group, it has declined more recently. When calculating within-group inertia, the deviations from the mean coordinate point are used, and the fact that the deviation, not original coordinate point, of A1 has declined has led to the decrease in its contribution. In other words, it is pointed out that the relative superiority of A1 in terms of educational status over other occupational groups

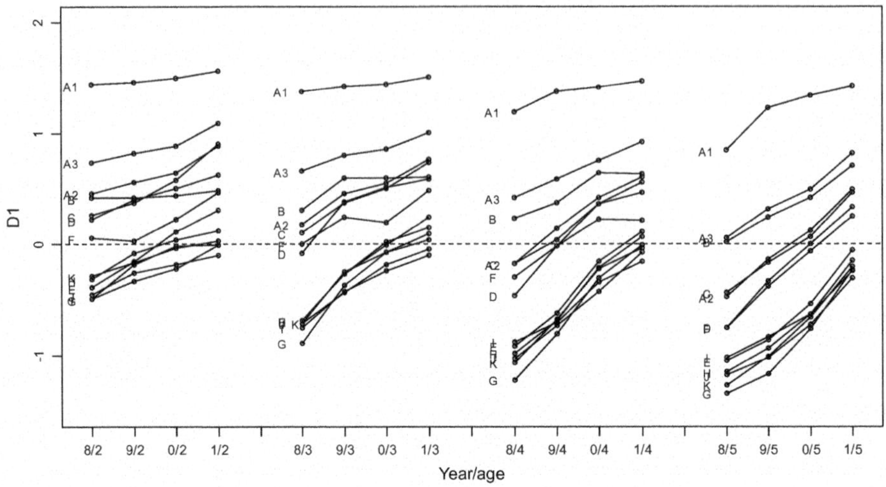

Fig. 6 Changes of the centroids of 13 occupational categories in dimension 1

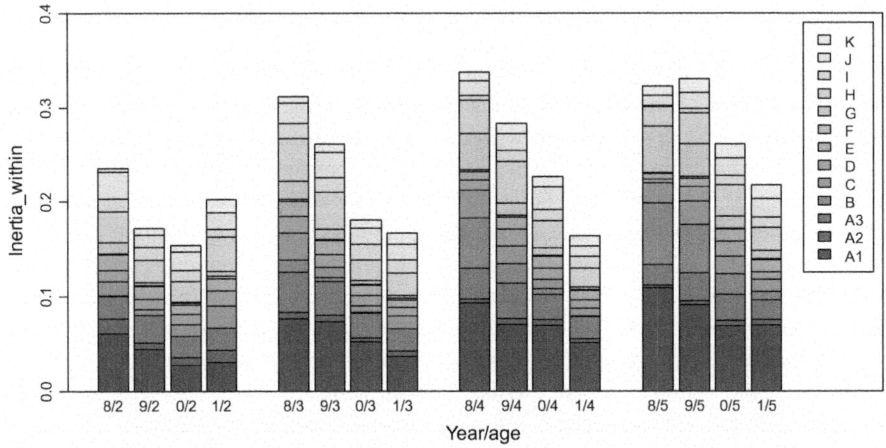

Fig. 7 Decomposition of within-group inertia into each occupational category in dimension 1

has gradually weakened against the background of an increase in the educational level of other groups as seen in Fig. 7.

Focusing on changes by age groups, with regard to the 20s, in which the overall contribution of dimension 1 increased again in 2010 as seen before, a larger increase is found in C (clerical) and H (manufacturing process): 0.0127–0.0237 and 0.0213–0.0372, respectively. In terms of the deviation (distance from the mean), the former is a positive value and the latter is a negative value, which indicates that occupational segregation has been strengthened again between clerical jobs for highly educated and manual jobs for poorly educated workers. For the 30s–50s age group, the contributions are declining in most occupations including A1, in which the decline is prominent in the 40s and 50s, especially in B (managerial). Whereas, in 1980, the contribution of B (40s: 0.0532, 50s: 0.0651) was the second largest after A1, they have declined to a very small contribution in 2010 (40s: 0.0013, 50s: 0.0084) as compared with other occupational groups. This change can be attributed not only to the fact that the educational superiority of managerial occupations has been reduced but also to the fact that the size of the working population has shrunk. Specifically, the population ratio has been changing from 6.9% (1980) to 1.9% (2010) in the 40s and from 8.6% (1980) to 3.7% (2010) in the 50s respectively. A managerial occupation was regarded as one of the career goals for highly educated non-manual workers (Hara 1981a). However, it is indicated that the decrease in the size (post) has resulted in the weakening of the importance of higher education.

(2) Changes in the dimension of gender

Figure 8 shows the changes in the centroids of each occupational category in the gender dimension (D2) for each age group. In the positive direction, the centroids of category I (transportation, machine operation), J (construction), F (protective), and B (managerial) are found. In spite of some variation, they are "male-dominated occupations" common among any year-age group. In the negative direction, the centroid of category A2 (professional and technical on personal service) is prominent, and it is recognized as a typical "female-dominated occupation" in any year-age group. The differences by age group are found in categories C (clerical) and E (service). The former has high negative values (clerical occupation is female-dominated) among the youth; while the latter has high negative values (service occupation is female-dominated) among the middle-aged and elderly. Regarding changes in time, there is a tendency for both to shift toward a higher negative direction (the direction toward being "female-dominant"). However, exceptionally, the negative value of C in their 20s is decreasing from 2000 to 2010.

According to the within-group inertia shown in Fig. 9, that of category C in the 20s has greatly decreased in 2010, which leads to the overall decline of contribution of dimension 2 in the age group. While the contribution of C was larger than any other occupational group in 2000, it plummeted in 2010, from the inertia of 0.0936–0.0311. The contribution of C in the 20s had been larger than those in other age groups between 1980 and 2000, which was a factor concerning the age difference in the contribution of dimension 2 as a whole. As was suggested from Fig. 3, there is a tendency for female workers to leave their jobs between their 20s and 30s. It would have the effect of loosening up gender segregation, which was significant in their 20s. That is, the segregation in clerical occupation (and eventually the whole) has been loosened by females who entered as clerical workers leaving their jobs because of marriage, childbirth, or child-rearing. It is also pointed out that the peak of such a phenomenon is 1990 during which the contribution of C is largest among those in the 20s. In 2010, gender segregation in clerical occupations was small even among the fledgling 20s. As a background for this, it is possible that female clerical workers are leaving jobs earlier. However, taking the trend of delayed marriage into

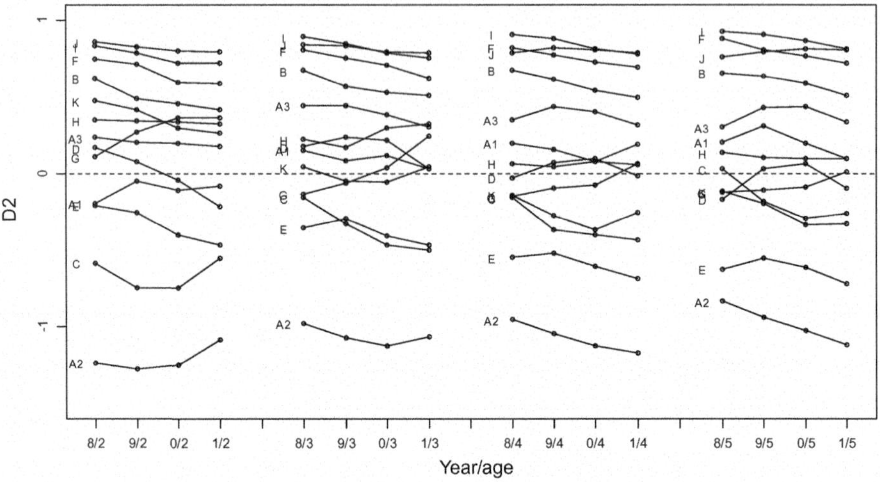

Fig. 8 Changes of centroids of 13 occupational categories in dimension 2

consideration, it is hard to believe that the number of women retiring for marriage and childbirth in their 20s is increasing. Therefore, this trend could be attributed to the fact that gender segregation is shrinking at the time of job entry.

On the other hand, the contribution of A2 in the 30s–50s has been rising, which leads to the increase of the overall contribution of dimension 2. This professional and technical with personal service occupation, including a large proportion of female workers such as nurses, welfare workers, and childminders, is characterized by marketable skills ("*te ni shoku*" in Japanese) for women ensured by professional qualifications. Even after leaving temporarily due to circumstances related to one's

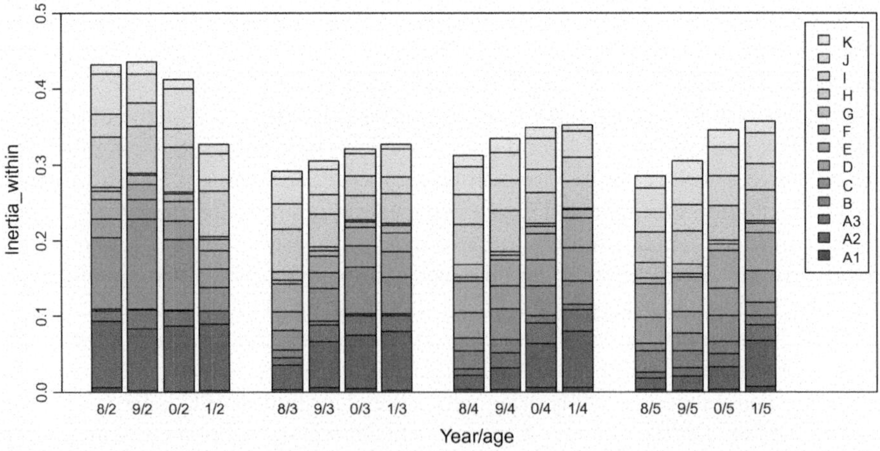

Fig. 9 Decomposition of within-group inertia into each occupational category in dimension 2

family, those who possess socially recognized qualifications would have an advantage during reemployment. Considering also that A2 has made consistently high contributions since 1980 in the 20s immediately after entering, the rising contribution in and beyond the 30s indicates that female workers of professional and technical with personal service occupations who continue to work or those who reenter the occupations after temporary leave are increasing. It can also be pointed out that the needs of care jobs, some of which are included in A2, has increased in recent years.

5 Summary and Conclusion

In this chapter, I have focused on and examined the time change of occupational segregation deriving the multidimensional structure of education and gender. The findings are that the segregation structure has not changed significantly in 30 years, while the importance of each element (dimension) has changed: the importance of segregation by educational qualification is declining and the relative importance of that by gender has increased. While many scholars expected the effect of education as an achieved status[11] on occupational attainment to become stronger as industrialization progresses (Treiman 1970; Hara 1981b), the results in this research do not support that expectation. The background to the decline in the influence of educational qualifications is as follows: first, the relative superiority of a "professional with advanced qualification," occupied by those with higher education, has been lost as the educational level of manual workers has risen. Second, the working "managerial" population, the career goal for the highly educated, has shrunk (the number of posts has decreased).

It has also been expected that the (direct) effect of the ascriptive status that an individual has from birth, such as sex, on occupational attainment diminishes as the achievement principle prevails. However, this expectation is also not supported. It is particularly important that even among those who have attained higher education, the gender segregation of occupations is as large as among others. Since higher education has an important role in acquiring specialized knowledge and skills that lead to job performance, there should not be gender differences in the role. The segregation in the *field of study* in higher education can be considered the background of gender segregation even among those with higher education (Charles and Bradley 2002; Naikakufu 2018: 131–132). In this chapter, the weight of "professional and technical with personal service" occupations mainly in medical and welfare fields was found to be increasing from 1980 to 2010 especially in the 30s–50s age groups. These occupations are regarded as those of market-relevant skills ("*te ni shoku*" in Japanese) for women. Since the possession of professional qualifications for these occupations facilitates re-employment of women after temporary leave, they accord with the career of women through their life course. On the other hand, it cannot be overlooked that the same thing also strengthens gender segregation. The segregation appeared smaller in the 30s–50s age group than in the 20s age group in the past because several female workers who had entered female-dominated occupations left the labor market for reasons such as marriage, childbirth, or child-rearing. In recent years, although continuous employment or reemployment (after temporary leave) of women has gradually increased, conventional male-dominated occupations have not become their new destination; female-dominated occupations continue to be their destination. It can be said that the quantitative expansion of female-dominated occupations while maintaining the gender segregation structure has resulted in the strengthening of the segregation.

Another noteworthy point is that in their 20s, individuals show differences from other age groups. Specifically, educational segregation had been growing in importance from 2000 to 2010, while, simultaneously, gender segregation had reduced importance. "Clerical" is the occupation common to the fluctuations of both. It was inferred that the tendency for males and people with higher education to enter clerical occupations has strengthened, based on the fact that the linkage with higher education has become stronger (again) and that the gender segregation has greatly decreased. In addition, the linkage with lower education has become stronger in "manufacturing process" occupations. It is said that non-regular employment of senior high school graduates has been increasing since the 2000s (Kosugi 2007: 43–47). The condition might be apparent among the workers of the manufacturing site. It will be necessary in future research to follow these changes brought about in recent youth, while also utilizing other data.[12]

Acknowledgements This research received grants from JSPS: 15K03818 and 18K01966. This paper revises the report "The Future on Education-based Society" (edited by Keiko Nakao) according to the former grant.

Appendix: Validity of 13 Occupational Categories

In Sect. 4.2, I tried to reproduce the configuration of the original classification (157 categories) as closely as possible by dividing "professional and technical" (A) into three categories. Here, 11 categories of JSOC and 13 categories in this study are compared from the viewpoint of the configuration in the dimensional space and the degree of explanation, supplementing the validity of using the latter.

First, Fig. 10 shows the configuration of the centroid (weighted by population) of each category. Note that these are the same year-age groups (4 groups) as in Fig. 4. While the centroids of categories A–K in JSOC cannot sufficiently express occupations located in the dimension 1 positive direction and the dimension 2 neagtive direction, classifying A into three categories A1–A3 can do so; that is, A1 (professional with advanced qualification) is representative of the former direction and A2 (professional and technical with personal service) is representative of the latter direction, respectively.

Next, in order to evaluate the degree of explanation, I calculate information loss (the smaller, more explanatory) when reducing the original 157 categories to 11 or 13 categories. The loss is expressed by within-group inertia based on the deviation

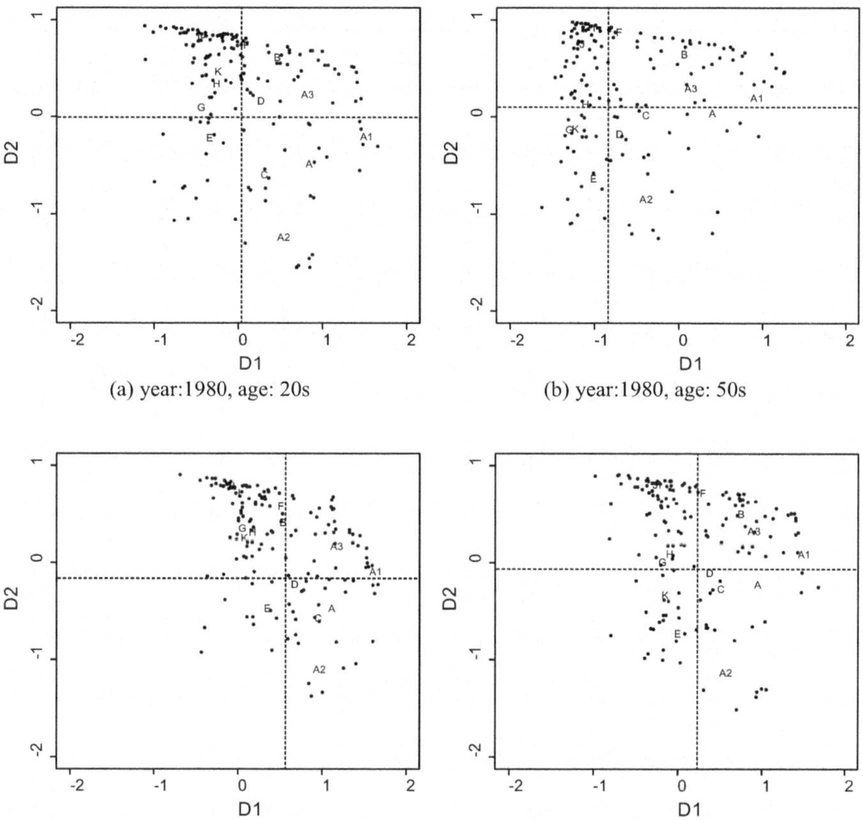

(a) year:1980, age: 20s

(b) year:1980, age: 50s

(c) year:2010, age: 20s

(d) year: 2010, age: 50s

Fig. 10 Centroids of the occupational categories in a two dimensional space

from the mean of each category, as the following Eq. (6):

$$\sum_{i=1}^{K} (x_{rit} - \overline{x}_{rkt})^2 \, m_{i\ \text{within\_}t} \tag{6}$$

Note that x_{rit} is the coordinate point in the rth dimension of each occupation ($i =$ 1, 2, …, K) included in the category k of the year-age group t, $\overline{x}_{rkt}$ is the mean of the category k, and $m_{i\ \text{within}\ t}$ is the ratio of each occupation i to the total population of the year-age group t.

Table 4 shows within-group inertia for each occupational category. As for the 11 categories, the extent of loss of "professional and technical" (A) is larger than that of the others both in dimension 1 and 2 (left table). When calculating the loss rate ("Loss %" in the table) by dividing the obtained total inertia by the inertia of the original 157 categories (Eq. 3), the loss to reach 50% is found in dimension 2. On the other hand, by classifying A into three categories, the inertia within categories A1–A3 greatly reduces in both dimensions 1 and 2 (right table). The loss rate of all 13 categories on the whole improved from 20.3 to 12.6% for dimension 1 and from 49.6 to 34.0% for dimension 2. Therefore, it can be said that this classification succeeded to a certain extent in suppressing the loss of information in the original classification.[13]

A comparison of total losses from different years by Table 5 shows that the improvement from 11 to 13 categories is larger in the recent years, especially for D2. For instance, among the 20s, the improvements of loss % for D2 are 13.0 points (42.2–29.2%) in 1980, 17.9 points (40.7–22.8%) in 1990, 18.6 points (39.2–20.7%) in 2000, and 21.8 points (46.0–24.2%) in 2010. This implies that it is more important in recent years than it was before to divide the category of "professional and technical" into three to represent gender segregation.

Notes

1. When the structure of prestige evaluation itself is the subject of study, it could be derived from the analysis. However, when applying the scale to another research, the prestige score is an external criterion.
2. Gender differences are also large when employment status is used as an indicator (see chapter "Occupational Segregation by Education and Gender in Japan: Focusing on Employment Status and Marital Status").

Table 4 Within group-inertia for each occupational category (summed up in all year-age groups)

11-categories				13-categories		
	D1	D2			D1	D2
A	0.3789	1.1505	→	A1	0.0092	0.0506
B	0.0094	0.0042		A2	0.0372	0.0811
C	0.0290	0.1305		A3	0.0369	0.1628
D	0.1270	0.4563				
E	0.0452	0.1850				
F	0.0155	0.0016				
G	0.0098	0.0646		(B - K ommited)		
H	0.0701	0.5714				
I	0.0183	0.0067				
J	0.0605	0.0064				
K	0.0147	0.1493				
Total	0.7785	2.7264		Total	0.4830	1.8704
Loss %	20.3%	49.6%		Loss %	12.6%	34.0%

Table 5 Total loss of inertia due to categorization for each year-age group

Year	Age group	D1				D2			
		11-Categories		13-Categories		11-Categories		13-Categories	
		Total	Loss %	Total	Loss %	Total	Loss %	Total	Loss %
1980	20s	0.0640	27.2	0.0449	19.1	0.1825	42.2	0.1263	29.2
	30s	0.0628	20.1	0.0449	14.4	0.1641	56.3	0.1351	46.3
	40s	0.0505	15.0	0.0307	9.1	0.1578	50.6	0.1400	44.9
	50s	0.0426	13.2	0.0220	6.8	0.1378	48.3	0.1246	43.7
1990	20s	0.0390	22.7	0.0232	13.5	0.1772	40.7	0.0994	22.8
	30s	0.0529	20.2	0.0319	12.2	0.1807	59.4	0.1199	39.4
	40s	0.531	18.7	0.0332	11.7	0.1752	52.3	0.1405	42.0
	50s	0.0527	16.0	0.0307	9.3	0.1523	50.0	0.1312	43.0
2000	20s	0.0382	24.7	0.0263	17.0	0.1620	39.2	0.0853	20.7
	30s	0.0409	22.7	0.0221	12.2	0.1674	52.1	0.0914	28.4
	40s	0.0541	23.9	0.0307	13.6	0.1928	55.3	0.1268	36.4
	50s	0.0536	20.5	0.0328	12.5	0.1740	50.3	0.1387	40.1
2010	20s	0.0470	23.3	0.0368	18.2	0.1502	46.0	0.0790	24.2
	30s	0.0354	21.2	0.0224	13.5	0.1628	49.8	0.0885	27.1
	40s	0.0398	24.3	0.0207	12.6	0.1953	55.3	0.1128	31.9
	50s	0.0519	23.8	0.0296	13.6	0.1943	54.4	0.1309	36.7

3. The results of analysis shown in this paper are based on the statistical deliverables by order-made summarization service of the "National Population Census" (Ministry of Internal Affairs and Communications) that the National Statistics Center provides under the rules of Statistics Act. Thus, it differs from the statistics published by the Ministry of Internal Affairs and Communications.
4. In the census, "specialized training college and miscellaneous school" are included in one of the school categories according to the entrance qualification and the term of study.
5. In the study by Hayashi et al. (2017) using the same data, correspondence analysis was applied to the contingency table of occupation × attribute (combination of sex, age, and education). We confirmed that the principal axes of the dimensional space at any time points were segregation of educational status and gender.
6. The program that performs correspondence analysis is "CA" function contained in the package "FactoMineR" (Lê et al 2008) of statistical software R (R Core Team 2017). Since the axis will be rotated and inertia will be recalculated based on the coordinate points as described later, coordinate points are not standardized.
7. The program for Procrustes rotation is the "procrustes" function contained in the package "vegan" (Oksanen et al. 2017) of statistical software R (R Core Team 2017).
8. Conversely, when comparing the magnitude of the contribution including the difference of mass of the group, it is better not to make such adjustments. Also, when adjusting, it is not necessary to set the sum of mass to "1" if the sum is equal (for example, "1/number of groups" when the sum of all the groups is 1).
9. Hayashi (2017) explores the multidimensional structure of occupation based on the subjective cognition of distance between a respondent him/herself and occupational groups and elicits "personal service with professional, technical skills" and "nonmanual in bureaucratic occupation" as the internal differentiation of female-dominated occupations. The former is

equivalent to "professional and technical with personal service" in this research. It is also equivalent to the "type-II professional" as described by Yamaguchi (2017). This category consists of "human service professionals" excluding those with high socioeconomic status such as university professors, physicians, and dentists (Yamaguchi 2017: 95).

10. It would be because the generation was educated during the fluctuating periods before and after World War II.
11. Note that there has been much controversy as to whether educational status is a *pure* achievement. For instance, it is strongly influenced by status of origin as Kajita (1981) described "ascribed achievement"; the educational status one has acquired changes into his/her ascription as Kajita (1981) described "achieved ascription"; it is a form of social closure with qualification limited to a specific status group (Collins 1979).
12. For implications from the view of educational sociology, refer to the study by Kondo (2017), which analyzed the relationship between education and occupation by the Gini segregation index using census data from 1960 to 2010.
13. In dimension 2, the inertia within the "manufacturing process" (H) and "sales" (D) is somewhat higher, suggesting that male-/female-dominated occupations are mixed among them. However, in this paper, priority is given to the division of "professional and technical"(A) because of the large number of occupational titles.

References

Bakker, B.F.M. 1993. A new measure of social status for men and women: The social distance scale. *The Netherlands' Journal of Social Science* 29: 113–129.

Benzécri, J.P. 1992. *Correspondence analysis handbook*. New York: Marcel Dekker.

Bijleveld, C. C. J. H., and E. van der Burg. 1998. Analysis of longitudinal categorical data using optimal scaling techniques. In *Longitudinal data analysis: Design, models and methods*, ed. C. C. J. H. Bijleveld and L. J. Th. van der Kamp (with A. Mooijaart, W. A. van der Kloot, R. van der Leeden, and E. van der Burg), 46–154. Sage publications.

Chan, T.W., and J.H. Goldthorpe. 2004. Is there a status order in contemporary British society? Evidence from the occupational structure of friendship. *European Sociological Review* 20 (5): 383–401.

Charles, M., and K. Bradley. 2002. Equal but separate? A cross-national study of sex segregation in higher education. *American Sociological Review* 67: 573–599.

Clausen, S.-E. 1998. *Applied correspondence analysis: An introduction*. Thousand Oaks: Sage Publications.

Collins, R. 1979. *The credential society: An historical sociology of education and stratification*. New York: Academic Press.

Genji, K. (ed.) 2018. *Koyou tayouka shakai ni okeru shakai teki chii no sokutei* (*Measurement of social status in employment diversified society*). Research report from the grant JSPS (in Japanese).

Gower, J.C., and G.B. Dijksterhuis. 2004. *Procrustes problems*. New York: Oxford University Press.

Greenacre, M.J. 1984. *Theory and applications of correspondence analysis*. London: Academic Press.

Greenacre, M.J. 1993. *Correspondence analysis in practice*. London: Academic Press.

Grusky, D. B., and J. B. Sørensen. 2001. "Are there big social classes?" In *Social stratification: Class, race, and gender in sociological perspective*, 2nd ed., ed. D. B. Grusky, 183–194. Boulder, CO: Westview Press.

Grusky, D. B. 2005. Foundations of Neo-Durkheimian Class Analysis. In *Approaches to Class Analysis*, ed. E. O. Wright, 51–81. Cambridge: Cambridge University Press.

Hara, J. 1981a. Shokugyou keireki no shakaigaku teki kenkyuu: toutatsu ten to kadai (A sociological study on occupational career: Findings and agendas). In *shokugyou no shakaigaku teki kenkyuu (sono 3)* (*Sociological studies on occupation [Part 3]*), 2–31. Koyou sokushin jigyou dan shokugyou kenkyuujo (Laboratory on Occupation in the Institute for Promotion of Employment) (in Japanese).

Hara, J. 1981b. "Kaisou kouzou ron" (Theory on stratification structure). In *Kiso Shakaigaku Dai 4 Kan: Shakai Kouzou* (*The basics of sociology Vol. 4: Social structure*), ed. S. Yasuda, T. Shiobara, K. Tominaga, and Y. Yoshida, 34–54. Tokyo: Toyo keizai shinpousha (in Japanese).

Hayashi, T. 2017. Shukan teki kyori hyoutei ni motozuku shokugyou ninchi jigen to shuudan bunka no sekishutsu: Reduced K-means cluster bunseki no tekiyou (Dimensions and group differentiation of occupational cognition derived by subjective distance evaluation: An application of reduced K-means cluster analysis). *Koudou Keiryou Gaku (The Japanese Journal of Behaviormetrics)* 44 (1): 73–84 (in Japanese).

Hayashi, T., Y. Saito, and K. Nakao. 2017. Kokusei chousa no order-made shuukei ni motozuku shokugyou data no riyou: sei, nenrei, gakureki ni motozuku shokugyou kan kyori no tajigen kouzou no sekishutu (The utilization of occupational data based on order-made summarization of census: Extract of multidimensional structure of occupational distance by sex, age, and education). *Report on the 90th annual meeting of Japan sociological society* (at Tokyo University) (in Japanese).

Kajita, T. 1981. Gyouseki shugi no naka no zokusei shugi ("Ascription" in the "Achievement Society"). *Shakaigaku Hyouron* (*Japanese Sociological Review*) 32-3: 70–87 (in Japanese).

Kimiyama, Y. 2005. *Correspondence bunseki no riyou hou* (*Usages of correspondence analysis*). Kanagawa: Data bunseki kenkyuujo (Laboratory of Data Analysis) (in Japanese).

Kondo, H. 2006. Idou hyou ni yoru shougyou teki chii shakudo no kousei: Ordination gihou no ouyou (Constructing occupational status scale from mobility tables: An application of ordination method). *Riron to Houhou (Sociological Theory and Methods)* 21 (2): 313–332 (in Japanese).

Kondo, H. 2017. Gini bunri shisuu wo mochiita gakureki—shokugyou kanren no bunseki (Analysis of the relationship between education and occupation using Gini-segregation indices). *Osaka Daigaku Daigakuin Ningenkagaku Kenkyuuka Kiyou (Bulletin of the Graduate School of Human Science, Osaka University)* 43: 85–102 (in Japanese).

Kosugi, R. 2007. Gakkou kara shokugyou eno ikou no hen'you (Transformation of transition from school to occupation). In *Freeter ni tairyuu suru wakamono tachi* (*Young people staying in "Freeter" [part-time workers]*), ed. Y. Hori, 31–99. Tokyo: Keiso Shobou (in Japanese).

Laumann, E.O., and L. Guttman. 1966. The relative associational contiguity of occupations in an urban setting. *American Sociological Review* 31 (2): 169–178.

Lê, S., J. Josse, and F. Husson. 2008. FactoMineR: An R package for multivariate analysis. *Journal of Statistical Software* 25 (1): 1–18.

Miwa, S. 2009. Shakai teki kakusa wo toraeru tameno micro-class approach: JGSS ruiseki data 2000–2006 ni yoru keiryou bunseki (Micro class approach to capture social disparities: Quantitative analysis of JGSS cumulative dataset 2000–2006). In *Nihon ban General Social Surveys kenkyuu ronbun shuu [8] JGSS de mita nihonjin no ishiki to koudou* (*Japanese general social surveys research series [8] Japanese consciousness and behavior in JGSS*), ed. Osaka University of Commerce and the Institute of Social Science at the University of Tokyo, 1–12. JGSS Research Center in Osaka University of Commerce (in Japanese).

Naikakufu. 2018. Heisei 30 Nen Ban Danjo Kyoudou Sankaku Hakusho (2018 Government white paper on gender equality). http://www.gender.go.jp/about_danjo/whitepaper/h30/zentai/index.html. Accessed 29 April 2019 (in Japanese).

Nakao, K., and J. Treas. 1994. Updating occupational prestige and socioeconomic scores: How the new measures measure up. *Sociological Methodology* 24: 1–72.

Okada, A., and T. Imaizumi. 1994. *Pasokon tajigen shakudo kouseihou* (*Multidimensional scaling using a personal computer*). Tokyo: Kyouritsu Shuppan (in Japanese).

Okada, A., and T. Imaizumi. 1997. Asymmetric multidimensional scaling of two-mode three-way proximities. *Journal of Classification* 14: 195–224.

Oksanen, J. F., G. Blanchet, M. Friendly, R. Kindt, P. Legendre, D. McGlinn, P. R. Minchin, R. B. O'Hara, G. L. Simpson, P. Solymos, M. H. H. Stevens, E. Szoecs, and H. Wagner. 2017. Vegan:

Community ecology package. R package version 2.4-3. https://CRAN.R-project.org/package=vegan. Accessed 21 Nov 2017
Prandy, K. 1990. The revised Cambridge scale of occupations. *Sociology* 24 (4): 629–655.
Prandy, K., and P.S. Lambert. 2003. Marriage, social distance and the social space: An alternative derivation and validation of the Cambridge scale. *Sociology* 37 (3): 397–411.
R Core Team. 2017. R: A language and environment for statistical computing. R Foundation for Statistical Computing, Vienna, Austria. https://www.R-project.org/. Accessed 21 Nov 2017
Rytina, S. 1992. Scaling the intergenerational continuity of occupation: Is occupational inheritance ascriptive after all? *American Journal of Sociology* 97 (6): 1658–1688.
Toubøl, J., and A.G. Larsen. 2017. Mapping the social class structure: From occupational mobility to social class categories using network analysis. *Sociology* 51 (6): 1257–1276.
Treiman, D.J. 1970. Industrialization and social stratification. *Sociological Inquiry* 40: 207–234.
Tsuzuki, K. (ed.). 1998. *Shokugyou hyouka no kouzou to shokugyou ishin score* (*The Structure of Occupational Evaluation and Occupational Prestige Scores*), 1995 SSM Survey Study Group (in Japanese).
Yamaguchi, K. 2017. *Hatarakikata no danjo fubyoudou: riron to jisshou bunseki (Gender inequality in the way of working: Theory and empirical analysis)*. Tokyo: Nihon Keizai Shinbun Publisher (in Japanese).

Occupational Segregation by Education and Gender in Japan: Focusing on Employment Status and Marital Status

Aya Watika

1 Introduction

Education plays a vital role in achieving occupational status in an industrialized society (Treiman 1970). The association between education and occupation is clear, even in recent years, when focusing on the multidimensional aspects of occupational segregation. According to Hayashi (chapter "Multidimensional Structure of Occupational Segregation by Education and Gender: Time Series Comparison in Japan from 1980 to 2010"), the effect of education on occupational status is evident when focusing on multidimensional association between educational level and occupational categories. Hayashi indicates that the link between gender and occupational categories has become important in recent years when compared to the association between educational level and occupation. Additionally, as Ohtsuki (chapter "Occupation Segregation among Immigrants in Japan: Using Custom-Made Census Data in 2010") demonstrates, ethnicity has been emerging as a "new" and important attribute in Japanese society. These results illustrate that although post-industrialization was widely spread among affluent societies, the effects of these attributes continue to be strong in Japan.

The international comparison of correspondence analysis (chapter "An International Comparison of Occupational Segregation by Gender: Through a Comparison of 14 Countries") measured occupational status by the International Standard Classification of Occupations (ISCO) major occupational groups and demonstrated that occupational gender segregation tended to be lower among highly educated workers in post-industrialized societies. However, there is strong gender segregation, even among those with degrees in Japan. This was clarified by conducting analyses

A. Watika (✉)
Division of Human Sciences, Ochanomizu University, 2-1-1 Otsuka, Bunkyo-Ku, Tokyo 112-8610, Japan
e-mail: wakita.aya@ocha.ac.jp

K. Nakao (ed.), *Multidimensional Aspects of Occupational Segregation*, Behaviormetrics: Quantitative Approaches to Human Behavior 18,
https://doi.org/10.1007/978-981-99-8513-5_3

dividing professionals and technicians/associate professionals into multiple occupational categories (chapter "Multidimensional Structure of Occupational Segregation by Education and Gender: Time Series Comparison in Japan from 1980 to 2010"; Yamaguchi 2019). It was thought that changes in the industrial structure strengthened the trends of female university graduates working not only as advanced professionals but also as interpersonal professionals or technicians/associate professionals, and of male and female junior college or technical college graduates working as technicians/associate professionals, clerks, and service/sales workers.

This chapter provides a detailed description of occupational segregation by gender and education, based on the characteristics of Japanese society. Therefore, the analysis in this chapter must consider important variables in contemporary Japanese society in addition to those used in the basic analysis conducted in chapter "Multidimensional Structure of Occupational Segregation by Education and Gender: Time Series Comparison in Japan from 1980 to 2010". Specifically, employment status and marital status were used as occupational and attribute variables, respectively.

1.1 Background

Education, rather than attributes (including gender), is considered to play an important role in achieving occupational status in industrialized societies (Treiman 1970), although studies of social stratification claimed cross-national similarity in social mobility patterns among capitalist, industrial societies (Featherman et al. 1975). Additionally, according to human capital theory, higher education has meritocratic power and weakens the disadvantages of ascription (Torche 2011), thus higher education among women should lead to an increase in their occupational status. Indeed, it has been argued that changes in the industrial structure have increased labor force participation among women, but the impact of *post-industrialization* on gender segregation is not so simple.

The concept of post-industrialization has several dimensions, such as the development of the knowledge industry, expansion of the service industry, feminization of labor, and meritocracy (Bell 1973; Esping-Andersen 1999). From the viewpoint of the effects on occupational gender segregation, post-industrialization resulted in service-sector expansion and economic rationalization (Charles and Grusky 2004). Charles and Grusky (2004) argued that post-industrialism strengthened horizontal occupational gender segregation between manual and non-manual workers and vertical occupational gender segregation among non-manual workers. They developed these hypotheses because routine non-manual occupations, especially those in the personal service industries, which had increased due to changes in the industrial structure, were traditionally perceived as domestic and feminine activities. These occupations attract female workers, including wives and mothers, who already had substantial domestic responsibilities. They examined these hypotheses about occupational gender segregation using data gathered from 10 market-based countries.

Oda et al. (2014) referenced Charles and Grusky (2004) and researched the impact of "post-industrialism" and gender egalitarianism on occupational gender segregation in Japan. Oda et al. hypothesized that gender egalitarianism weakened vertical occupational segregation in non-manual occupations even in Japan. They also hypothesized that post-industrialism strengthened horizontal occupational segregation, *weakened* vertical occupational segregation within non-manual occupations, and strengthened vertical occupational segregation within manual occupations. The analysis by Oda et al., which compared prefectures in Japan, tested this hypothesis. It should be noted that the horizontal segregation referred to here is the gender segregation between non-manual and manual occupations. Vertical segregation is the difference in the socio-economic index (assigned to each unit occupational group) by gender among non-manual workers and among manual workers. Charles and Grusky (2004) stated that the expansion of the service industry increased the number of women in lower non-manual occupations, so post-industrialism strengthened segregation by gender within non-manual occupations. However, Oda et al. (2014) stated that post-industrialism has not advanced much at a national level in Japanese society, which has resulted in women with higher human capital advancing to upper non-manual occupations in prefectures where post-industrialism has relatively advanced, thereby weakening gender segregation among non-manual occupations. Indeed, the level of post-industrialization is thought to be lower in Japan than in other affluent societies. This can be seen from Table 1, which depicts the percentage of manual workers (including "Crafts and related trades workers," "Plant and machine operators and assemblers," and "Elementary occupations" as ISCO major occupational groups) according to the census data analyzed in chapter "An International Comparison of Occupational Segregation by Gender: Through a Comparison of 14 Countries", of 12 countries around the year 2000. The percentage of manual workers in Japan remains high and the percentage of upper non-manual workers in Japan is lower compared to other countries, such as the UK, the U.S., the Netherlands, and France.

Changes in the industrial structure in Japan have so far been unlikely to weaken overall occupational gender segregation according to analyses using smaller occupational classifications that distinguish between various occupations within non-manual and manual occupations. Uchikoshi and Mugiyama (2020), which used micro occupational category (unit group) and measured the dissimilarity index, established that occupational gender segregation had a slightly decreasing tendency in Japan from

Table 1 Percentages of manual workers

China	17.7%	India	30.0%
UK	20.5%	Poland	33.2%
Thailand	24.1%	Japan	34.0%
Vietnam	24.3%	Malaysia	36.4%
U.S.	24.5%	Italy	37.6%
Netherlands	25.1%	Hungary	38.8%
France	28.2%	Philippines	43.0%

1980 to 2005. They showed that changes in the distribution of occupational categories (peripheral frequencies) offset the effects of changes in gender composition, although changes in the gender composition of occupational categories lowered the level of occupational gender segregation. As mentioned above, Hayashi's analysis of education/gender and occupational categories (unit group of population census) showed that gender factors became relatively more important for occupational segregation (chapter "Multidimensional Structure of Occupational Segregation by Education and Gender: Time Series Comparison in Japan from 1980 to 2010").

The results of Oda et al. (2014) suggested that post-industrialism led to women with high human capital advancing into upper non-manual occupations, and those with less flowing into lower manual occupations. Their analysis focused on differences between prefectures, so it was thought that several regions in Japan experienced such effects of post-industrialism. In contrast, the international comparative analysis in chapter "An International Comparison of Occupational Segregation by Gender: Through a Comparison of 14 Countries" showed that there was relatively strong occupational gender segregation even among workers with high human capital in Japanese society as a whole.

Therefore, among developed countries, Japan is considered to have relatively strong occupational gender segregation according to chapter "An International Comparison of Occupational Segregation by Gender: Through a Comparison of 14 Countries", but there are occupational variables that should be further considered when investigating segregation in Japan. Given Japan's employment systems and practices, occupational gender segregation is thought to be associated with various occupational variables other than occupational categories, such as employment status, firm size, and industry. There are clear differences in status by gender in the Japanese labor market, particularly in wages, employment status, and proportion of managers. Therefore, using only occupational categories as occupational status is insufficient to understand the low occupational status of Japanese women. Including employment status (regular/non-regular workers) in the analysis is especially important. The "deregulation" and "fluidization of employment" in labor policy since the mid-1980s have diversified the employment status of workers. Kambayashi (2017) argued that declining self-employment led to an increase in non-regular employment. In 2019, 38.5% of employees were non-regular workers, including part-time, temporary, dispatch, and contract workers (Statistics Bureau of Japan 2020). In 1984, this proportion was 15.3%. Figure 1 shows the percentage of female employees by occupation (minor classification) and the percentage of regular employees among workers, according to anonymized data from the 2007 Employment Status Survey. The letters in the figure correspond to those in Table 2. Among occupations with many female employees, the percentage of non-regular employment is high.

Non-regular workers tend to obtain fewer social resources, such as wages, compared to regular workers, and it is difficult for individuals to move from non-regular employment to regular employment. Thus, non-regular employment should be considered a social stratum in contemporary Japanese society (Taroumaru 2009). The wage gap due to the employment status is an especially serious issue. Suzuki (2018) examined the dual labor market in Japan. According to her, there are two

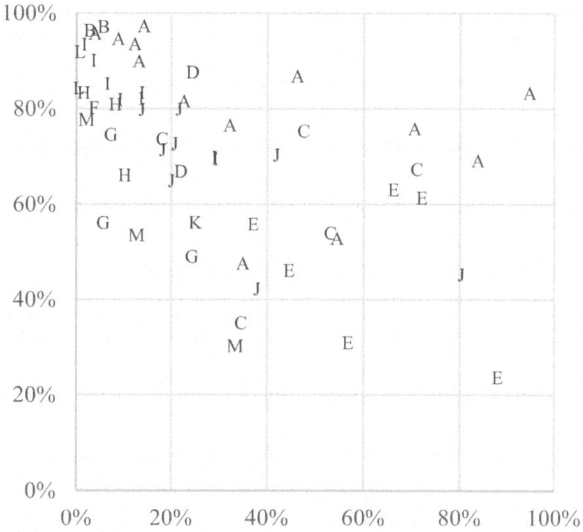

Fig. 1 Percentages of female workers (X-Axis) and percentages of regular workers (Y Axis)

Table 2 Major occupational classification

A	Professional and engineering workers
B	Administrative and managerial workers
C	Clerical workers
D	Sales workers
E	Service workers
F	Security workers
G	Agriculture, forestry, and fishery workers
H	Transport workers
I	Manufacturing process workers (metal, machines, chemical products)
J	Manufacturing process workers (food, fiber)
K	Manufacturing process workers (others)
L	Machine operation and construction workers
M	Carrying, cleaning, packaging, mining, and related workers

segments in the wage determination system in Japan, and employment status is an important determinant of the segment in which a worker is included. Low wages for women, especially married women, are closely related to their employment status. Arita (2016) revealed that in comparison with South Korea and Taiwan, employment status and firm size are important determinants of wages in Japan. Including employment status is necessary to analyze occupational gender segregation based on the characteristics of Japanese society.

Differences in the careers of men and women also ought to be examined to understand the low status of women with domestic responsibilities in the labor market. A common feature among industrialized East Asian countries is that married women have a low position in the labor market. In Japan, they tend to experience interruptions in their careers and re-enter the labor market as non-regular workers under characteristic employment practices (Brinton 2001). "Male-breadwinner"-type labor and welfare policies have supported the interruption of married women's occupational careers due to life events, such as marriage and birth of first child (Osawa 2007). The argument that policy interventions for gender equality are associated with occupational gender segregation has been made along with international comparisons (Chang 2000). Some recent studies have addressed gender segregation through the extent of having paid work throughout one's life course and occupations (Guinea-Martin et al. 2018). However, the impact of one's life course is considered particularly important in Japan, even when focusing exclusively on segregation of occupational status.[1] Analyses that include not only employment status variables but also variables related to the career and life course of women need to be conducted to understand the characteristics of occupational gender segregation in Japan.

Based on the this background, this chapter explores in detail the characteristics of occupational segregation by education and gender in contemporary Japanese society. The relationships between occupational status measured by occupational category (occupational type) and employment status, and attribute variables (age group, gender, and educational level) were clarified. Additionally, marital status (unmarried/married) was analyzed as an attribute variable to investigate the effect of gender-based occupational career differences on occupational segregation. Occupational careers cannot be treated as variables, but their effects can be partly understood by examining gender differences in the effects of marital status. If interruptions in the careers of women caused by life events strengthen occupational gender segregation, then there should be greater levels of segregation between married men and women than between unmarried men and women.

2 Data and Method

The non-linear association between attributes/education and occupational status was analyzed using anonymized data from the 1992, 1997, 2002, and 2007 Employment Status Surveys, which were provided by the National Statistics Center. The sample sizes were approximately 820,000, 800,000, 750,000, and 760,000, respectively. Additionally, aggregated data from the Employment Status Surveys of 2012 and 2017 published on the website "e-stat" of the Statistics Bureau of Japan were used (Statistics Bureau of Japan 2020). The attribute variables of age group (20s/30s/40s/50s), gender (male/female), educational level (junior high school/high school/junior college or technical college/university), and marital status (unmarried/married) were used as the row variables. The occupational variables of occupational category (major and minor groups of the Japanese Standard Classification of Occupation (JSCO) and employment status (regular employment/non-regular employment/self-employment or family employee) were used as the column variables.

3 Results

3.1 Occupation (Major Classification) and Employment Status from 1992 to 2017

In the first analysis, the column (occupational) variable was a combination of occupation (major classification) and employment status. Changes in occupational segregation by education and gender over time were shown by comparing the correspondence analysis based on each year's data. Figure 2 illustrates the contribution ratios of the dimensions obtained by the correspondence analysis for each year. A two-dimensional solution was adopted for all years. The contributions of these two dimensions have increased in recent years.

Figure 3 demonstrates the configuration of the attributes from 1992 to 2017. The first dimension (horizontal axis) and second dimension (vertical axis) can be interpreted as the "gender dimension" and "educational dimension," respectively. Figure 2 shows that the contribution of the first dimension increased since the 2000s. These results reveal that nowadays gender is a more important factor in determining occupational status than educational level as shown in chapter "Multidimensional Structure of Occupational Segregation by Education and Gender: Time Series Comparison in Japan from 1980 to 2010", even when employment status is included in the analysis. Moreover, Fig. 3 reveals that men in their twenties have drawn closer to women since the 2000s. This indicates that younger men are more likely to have jobs similar to women than older men do. However, this analysis did not confirm strong occupational gender segregation among highly educated workers in Japan, unlike the results in chapter "An International Comparison of Occupational Segregation by Gender: Through a Comparison of 14 Countries".[2]

Fig. 2 Contribution ratios of dimensions (Correspondence analysis in sect. 3.1)

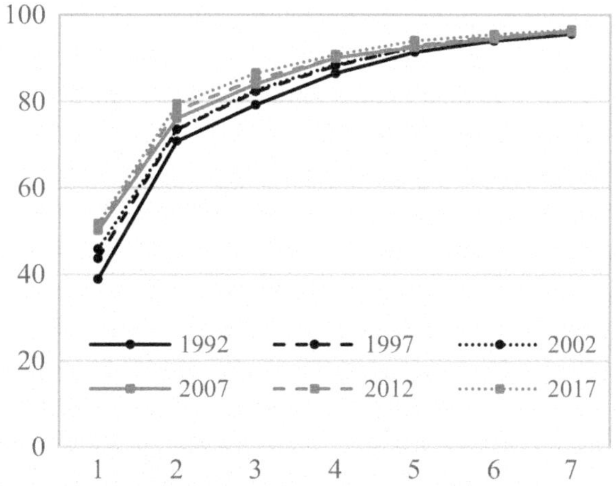

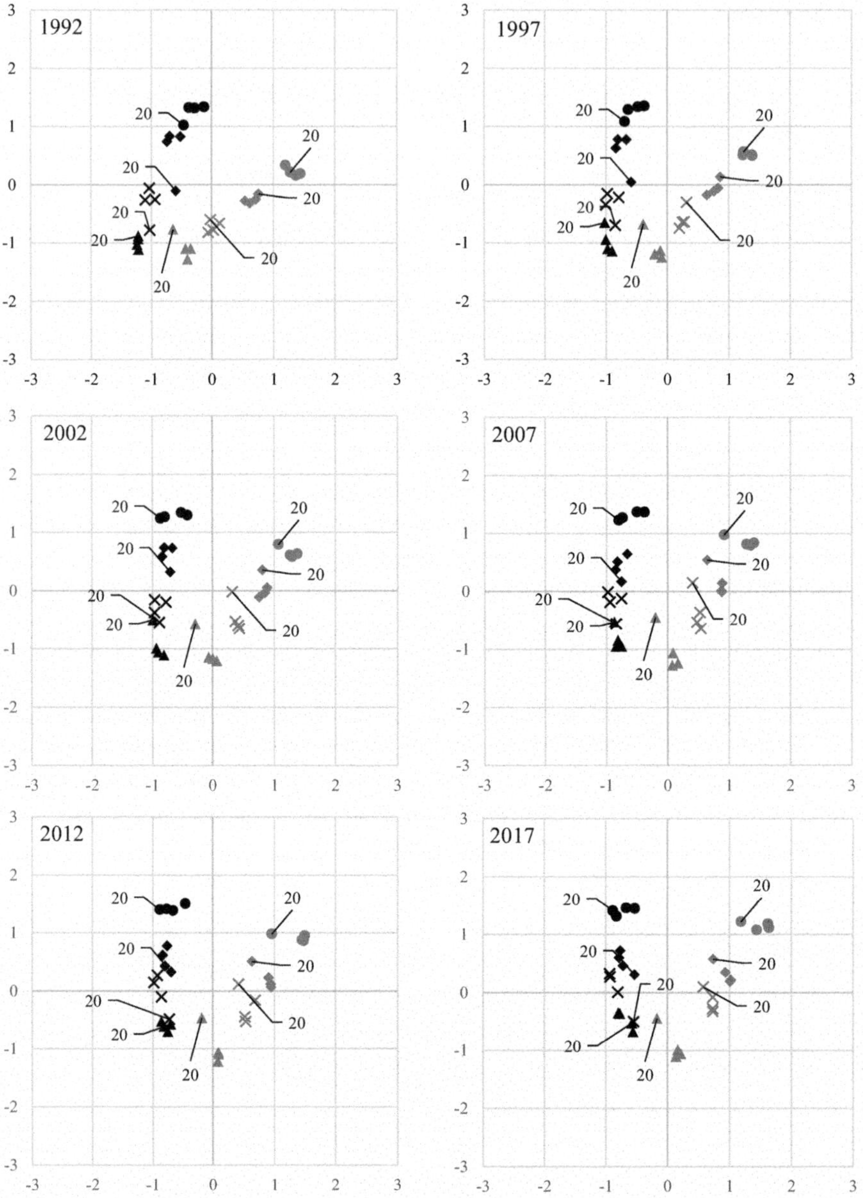

Black: Women, Gray: Men

●: Junior high school, ◆: High school, ×: Junior college, ▲: University 20: workers in their 20s

Fig. 3 Configuration of attributes (Correspondence analysis in sect. 3.1)

Figure 4 shows the configuration of the occupational categories (major occupational classification) and employment status.[3] The letters in the figure correspond to the occupational categories (major occupational groups), as shown in Table 2. The correspondence between Figs. 3 and 4 shows that there is a strong association between gender and employment status. Non-regular employees are located on the left (female) side of the figures, and regular employees are located on the right (male). Occupational segregation from the perspective of employment status is mainly caused by gender. There are several exceptions to this, however. "Transport workers (H)" and "machine operation and construction workers (L)" define occupations that men can find easily even for non-regular employment. These occupations are associated with the stereotype of "masculine" jobs. Meanwhile, the positions of regular "professional and engineering workers (A)" and "clerical workers (C)" are closer to highly educated women and young men than to other regular employees. This is probably because interpersonal professional and technical occupations include many female workers. Clerical occupations have many female employees, but highly educated men who will become managers in the future also start their occupational careers from clerical occupations. Therefore, this occupational category includes highly educated young men and women. Regularly employed "service workers (1E)" are located near moderately educated women.

Has a time-series change been observed in occupational segregation? From the 1990s to the 2010s, there were no major changes in the occupational configuration. Therefore, it is thought that the structure of occupational status has not vastly changed from the perspective of its relationship with education and attributes (age group and gender). There were no drastic time-series changes in the correspondence between attribute variables (Fig. 3) and occupational variables (Fig. 4). However, it should be noted that non-regular employees in professional and technical occupations have drawn closer to non-regular employees in other occupations since 2007. In the past, non-regular professionals and technicians (2A) corresponded to relatively highly educated women. However, these professions have not matched with highly educated individuals in recent years. The tendency of the status of non-regular employees (excluding "transport workers (2H)" and "machine operation and construction workers (2L)") to correspond to women with a low educational background has become clear in recent years. From the analysis results, it can be surmised that even when considering employment status as occupational status, there is no substantial change in the structure of the relationship between occupation and attributes/education. As shown in Fig. 2, occupational gender segregation has simply increased from the 1990s to the 2010s in Japanese society.

Charles and Grusky (2004) and Oda et al. (2014) suggested that the increase in interpersonal professionals and technicians caused by post-industrialism was a factor that expanded (horizontal) occupational gender segregation. The tendency of highly educated women to enter these occupations is reflected in the results above.[4] Additionally, the expansion of non-regular employment may contribute to strengthening occupational gender segregation. As mentioned above, the correspondence of the configurations (Figs. 3 and 4) reveals that not only the segregation of occupational groups (occupational type) by gender but also the segregation of employment

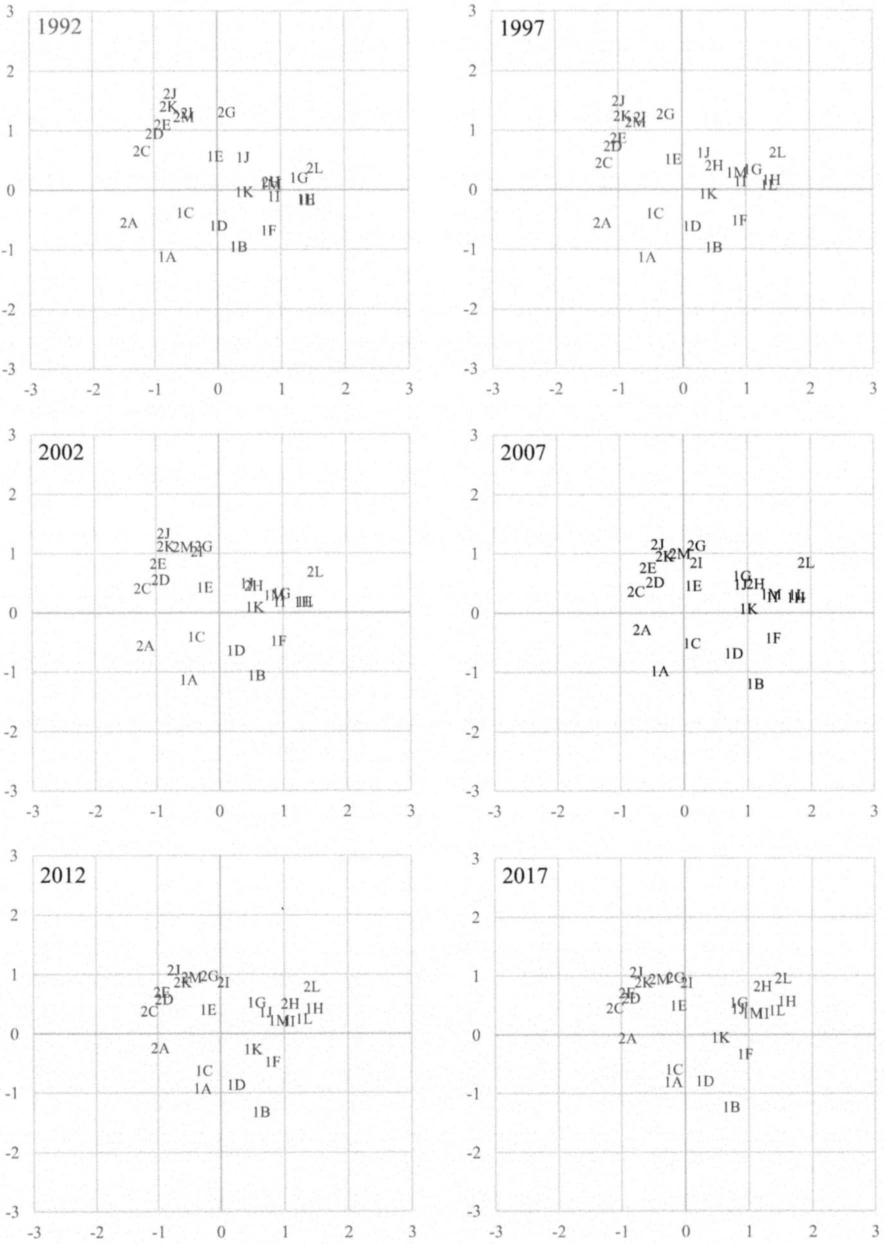

1: Regular workers, 2: Non-regular workers

Fig. 4 Configuration of occupational status (Correspondence analysis in sect. 3.1)

Fig. 5 Odds ratios for upper white-collar occupations (university vs. high school, controlled by age group)

status is large. In particular, the tendency for non-regular employment to correspond to women with low levels of education has become clearer in recent years.

There is a possibility that occupational segregation by educational level among women will increase if highly educated women enter upper non-manual occupations. Figure 5 shows the odds ratios for upper white-collar occupations (professional/technical occupations and managerial occupations, excluding non-regular employees) by age-controlled logistic regression analysis using data up to 2007. The logistic regression analysis included unemployed individuals.

In the past, educational background had a large effect on the ease with which women could become upper white-collar workers. However, the odds ratios of university education among women rapidly decreased from the 1990s to the 2000s, reaching the level of men in 2007. These results seem to indicate that the value of university education as human capital is, on the contrary, decreasing among women. This may be related to the indication that the value of a degree is lost due to growth in higher education (Horowitz 2018). It has been indicated in Japan, as well, that not only whether individuals received higher education but also whether they were admitted to prestigious universities is important in measuring educational level (Hirasawa 2011). Women's university enrollment rate, which was less than 20% in the early 1990s, rose sharply, reaching 50% in the late 2010s, approaching that of men. Therefore, the rarity of university education among women decreased to the same level as that in men. Thus, it can no longer be said that the disparities by human capital among women are widening regarding the ease of becoming regular employees in professional/technical occupations.

However, the difference in employment status depending on educational level can be cited as an aspect of the effect of high levels of education in women from the 1990s to the 2010s. Figure 6 shows the odds ratio of university education (vs. high school education) for regular employees, which was calculated by age-group-controlled logistic regression analysis.

The odds ratios for both men and women were not large but increased slightly. As non-regular employment expanded throughout society, a link between educational level and employment status was observed because university education was less likely to result in non-regular employment.

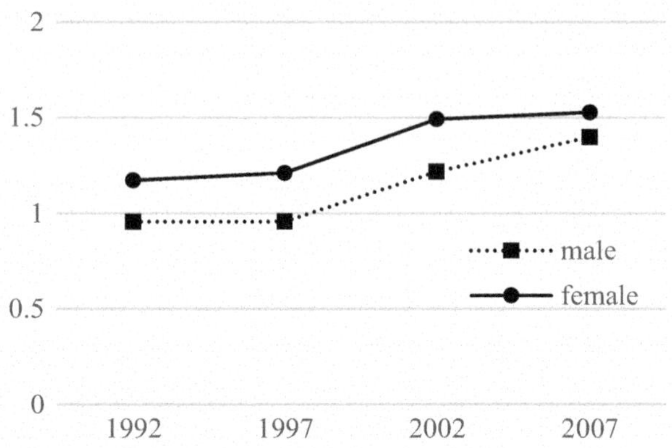

Fig. 6 Odds ratios for regular employment status (university vs. high school, controlled by age group)

3.2 Occupation (Minor Classification) and Employment Status in 2002 and 2007

The anonymized data of the 2002 and 2007 Employment Status Surveys included the JSCO minor groups (64 categories) as variables. These two datasets were used to show the results of the correspondence analysis using minor occupational classification and employment status as occupational status.

As shown in Fig. 7, the first dimension (horizontal axis) and second dimension (vertical axis) can be interpreted as the axes for gender and educational level, respectively, in these analyses as well. The contribution ratios of the first and second dimensions in 2002 were 37.57% and 23.33%, respectively. In 2007, they were 40.44% and 24.16%, respectively. The attribute configuration did not differ largely from the analyses using the major occupational group (Fig. 3); however, occupational gender segregation among highly educated individuals was more prominent.

Occupational configurations for regular and non-regular employee are shown separately to improve the readability of the figure (Fig. 8).[5] The letters representing occupations correspond to the major occupational classifications listed in Table 2. The clear tendency of men to be engaged in regular employment and women in non-regular employment, in addition to separation by occupation, is evident, even when using the minor group. However, there are some exceptions. Professional/technical occupations related to medical and welfare fields (nurses including assistant nurses and social welfare specialist professionals), musicians, stage designers, some clerical occupations (accountancy clerks, office appliance operators), many service occupations, and fiber product manufacturing workers are occupations that are dominated by women with low or moderate educational backgrounds, even if they are regular employees. Non-regular employees, such as musicians, stage designers, teachers, and other specialist professionals (including librarians and private tutors), are located close to highly educated women. Those occupations with the stereotype of being “feminine” occupations are easy for women to find as regular employment.[6] Meanwhile, occupations that men dominate even if it is non-regular employment include

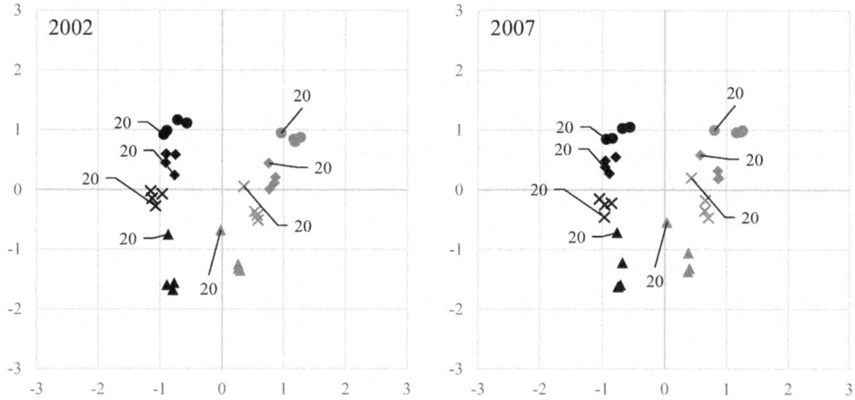

Black: Women, Gray: Men

•: Junior high school, ◆: High school, ×: Junior college, ▲: University 20: workers in their 20s

Fig. 7 Configuration of attributes (Correspondence Analysis in sect. 3.2)

occupations such as construction and civil engineering workers, stationary and construction machinery operators, transportation machinery assembly/maintenance and repair workers, and motor vehicle drivers. Occupations that men with moderate to high educational backgrounds tend to enter, even if they are non-regular employees, include security workers and engineers. These are professions with the stereotype of being "masculine." No major time-series change was seen between 2002 and 2007, and the configurations for both occupations and attributes were stable.

3.3 The Effects of Marital Status from 1992 to 2007

Finally, marital status was added as an attribute variable for the data in 1992 and 2007, and its relationship with occupational status (major occupational group and employment status) was investigated.[7] The contribution ratios of the first dimension were 36.83% and 47.71% in the 1992 and 2007 data, respectively. The contributions of the second dimension were 31.29% and 25.88%, respectively.

Figure 9 shows the attribute configurations by gender. Figure 10 illustrates the configurations of occupational status. When marital status was added as a new attribute variable, married men had higher first dimension values than unmarried men, while married women had lower first dimension values. In other words, there was a greater difference in first-dimension values by gender among married workers than among unmarried workers. As in the previous analyses, the first dimension (X-axis) can be interpreted as the gender axis and the second dimension (Y-axis) can be interpreted as the educational level axis. Therefore, the results can be interpreted as showing that occupational gender segregation is greater among married workers

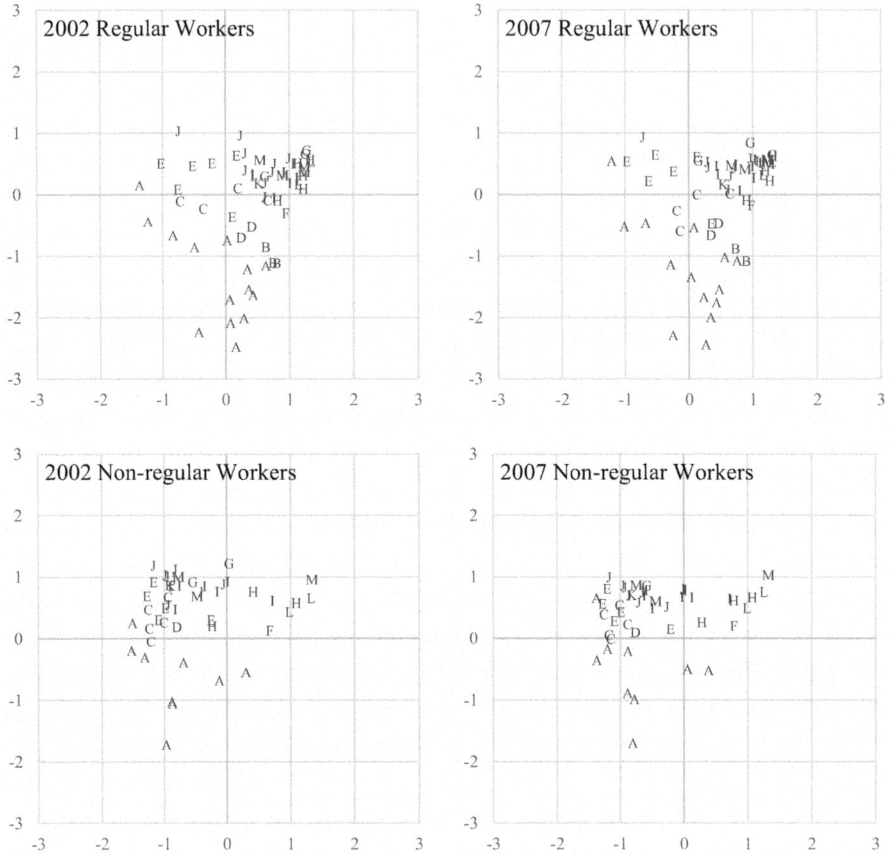

Fig. 8 Configuration of occupational status (Correspondence analysis in Sect. 3.2)

than among unmarried workers. On the other hand, occupational gender segregation was observed even among unmarried individuals and at any educational level. It is thought that occupational gender segregation, which is also present among unmarried individuals, is strengthened by marriage. Interruptions in careers due to life events are known to change the position of women in the labor market, especially when they shift to unemployment or non-regular employment after the birth of their first child. However, the result suggests that occupational gender segregation is prevalent even in the absence of life events. It is also not the case that the value of one's educational level is lost, even among married women.

Interestingly, the analyses in 3.1 showed that the configuration of men in their 20s tended to be more similar to that of women since the 2000s; however, according to the 2007 analysis in this section, this tendency was only seen in unmarried men. Married men, even younger individuals, did not occupy a position similar to women. Why is this the case? Women are known to lose their position in the labor market because

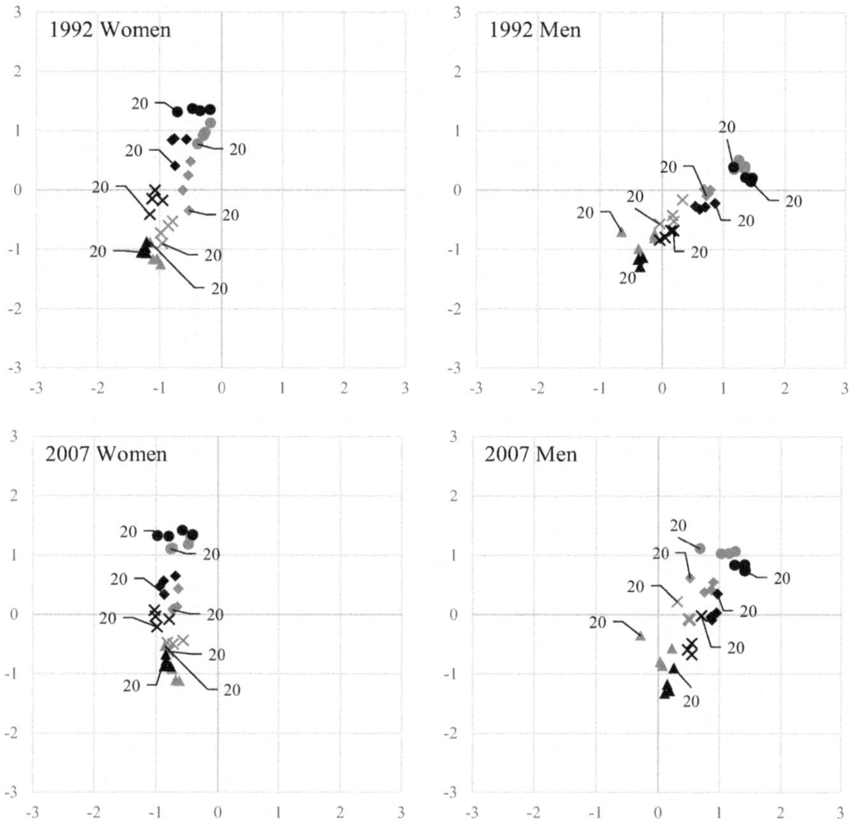

Black: Married, Gray: Unmarried

•: Junior high school, ◆: High school, ×: Junior college, ▲: University 20: workers in their 20s

Fig. 9 Configuration of attributes (Correspondence analysis in sect. 3.3)

of life events such as marriage, but this is not the case for men. It is thought that these results were obtained because men whose occupational status was "close to that of women" did not marry. The reason this tendency is predominant in recent years is thought to be that the scale of non-regular employment and of the professional/technical occupations related to medical and welfare fields and service occupations has increased, and the number of men with these occupational status has escalated. Occupational status relatively "close to women" features low wages. Men "close to women" in terms of employment status and occupational type may have lower wages and tend to remain unmarried. Figure 11 shows the percentage of female employees by occupation (minor occupational classification) and the average annual

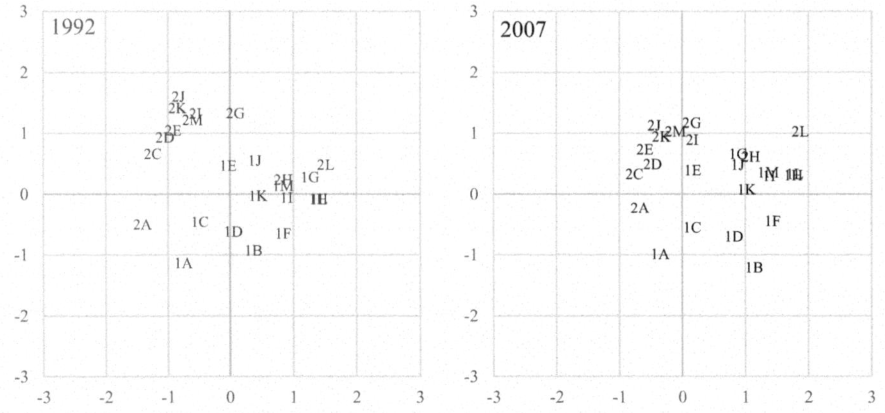

1: Regular workers; 2: Non-regular workers

Fig. 10 Configuration of occupational status (Correspondence analysis in sect. 3.3)

salary of regular employees according to anonymized data from the 2007 Employment Status Survey. The letters in the figure correspond to those in Table 2. Occupations with many female employees tend to have low wages, even when they are regular employees or in professional/technical occupations (A).

Will the change in the occupational status of unmarried men in their 20s "becoming closer to that of women" weaken occupational gender segregation in the long term? Whether the age group effect is a generational effect or an age effect cannot be determined unless we examine the position of the young cohorts in this study in the

Fig. 11 Percentages of female workers (X-axis) and average annual income (10,000 yen, regular workers, Y-axis)

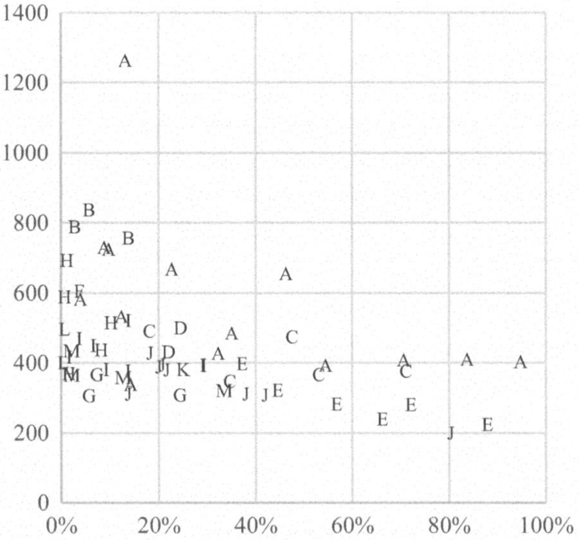

labor market 10 or 20 years later. In fact, the characteristics of unmarried workers in their 20s in the 1992 and 1997 data do not match the characteristics of unmarried workers in their 30s in 2002 and 2007, respectively.[8] It is not necessarily the case that the position of unmarried men in their 20s in 2007 would continue in 2017. However, the number of unmarried people has been rapidly increasing in Japanese society, and attention needs to be paid to the occupational status of unmarried individuals. The proportion of unmarried men aged 30–34 years was 28.2% in 1985 but increased to 47.1% in 2015. The proportion of unmarried women has also increased from 10.4 to 34.6% (Statistics Bureau of Japan 2016).

Large vertical segregation is one of the reasons why the overall perspective on the reduction of gender segregation is not optimistic. Although not covered in this chapter, vertical gender segregation, even in relatively small occupational categories, remains large. Figure 12 shows the percentage of female employees by occupational type (minor occupational classification) and the gender income gap between regular employees (male workers' income = 100%), which were calculated from the anonymized data of the 2007 Employment Status Survey. There is an overall gender difference in annual income, even among regular employees. The gender wage gap is mitigated in some professional/technical (A) and service (E) occupations, which have an extremely large number of female workers; however, wage levels are not high in these occupations (Fig. 11).

Fig. 12 Percentages of female workers (X-axis) and gender income gap (Y-axis, male workers' income = 100%)

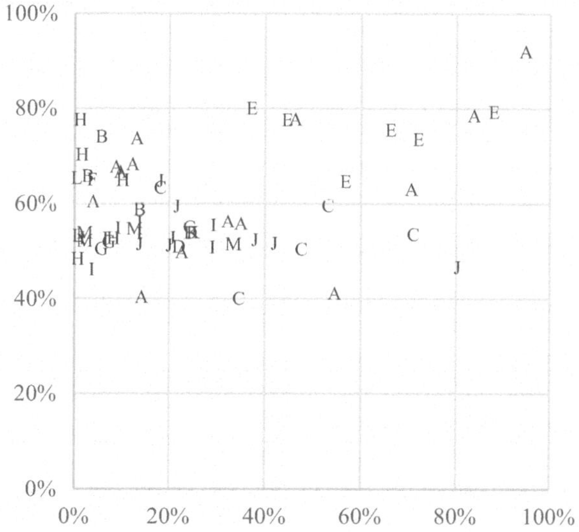

4 Discussion

The analysis results of this chapter are summarized, and an outlook on gender segregation in Japan is provided here. In this chapter, analyses including employment status as an occupational status are conducted (Sect. 3.1). This analysis indicates that an examination of employment status is crucial to understanding occupational gender segregation in Japan. Furthermore, analyses using employment status showed that occupational structure was maintained in relation to attributes and gender factors were relatively important for occupational segregation after the 2000s. It cannot be said that occupational gender segregation has weakened in Japan due to women's improved access to higher education in the post-industrialization scenario. Instead, the tendency of highly educated women to become regular employees in professional/technical occupations has decreased to around the same level as that of men from the 1990s to the 2000s. The university entrance rate for both sexes in Japan was only approximately 50%, but the value of women's degrees decreased. It cannot be said that the extent of post-industrialization in Japanese society is high compared to that in other developed countries. Therefore, it is thought that transitions in the industrial structure have not caught up with the popularization of higher education among men and women.

Thereafter, the analysis, using the minor occupational group, demonstrated that occupational segregation by gender is strong, even among highly educated individuals, as indicated in chapter "Multidimensional Structure of Occupational Segregation by Education and Gender: Time Series Comparison in Japan from 1980 to 2010" (Sect. 3.2). Professional/technical occupations with numerous female regular employees are limited to occupations with "feminine" stereotypes. Women's higher education is less linked to regular employment in professional/technical professions other than in "feminine" occupations. Gender differences in selecting academic majors in higher education may be an influencing factor (Horowitz 2018); there is a particularly large gender difference between academic majors in Japan (Cabinet Office 2019). Wages for occupations with a large female worker population tend to be low (Magnusson 2009), while "masculine" traits are considered necessary for success in occupations with high prestige (Cejka and Eagly 1999). Furthermore, occupations requiring traits that men are considered to have, pay higher wages (e.g., Levanon and Grusky 2016). Women's tendency to concentrate on non-regular employment is also a major problem in Japan.

The careers of married women are affected by the problem of them being considered suitable only for non-regular employment. In this chapter, an analysis that included marital status was conducted to investigate the differences in women's careers from those of men (Sect. 3.3). Gender and educational level had pronounced effects on the correspondence between occupational status and attributes, even when marital status was considered. There were high levels of segregation even between single men and women. Marriage does play a role in strengthening the influence of gender, but occupational gender segregation cannot be explained only by the interruption of female employment due to marriage and childbirth. Segregation persists from the time they are unmarried. This is consistent with the indication that women have been at a disadvantage in the workplace since their early careers and struggle to remain hopeful about their future careers (Iwata et al. 2015; Ohtsuki 2015). Furthermore, after the economic bubble burst in the 1990s, it has been indicated that non-regular employment has spread not only among married women but also among young women (Hirata 2011; Iwai 2011). The analysis focusing on employment status and marital status in this chapter indicated the large effect of life events on careers, but also that women are segregated into unfavorable occupational statuses (occupational type and employment status) since the beginning of their careers.

Due to changes in the 2000s, unmarried men in their 20s began to occupy a position in the labor market that was close to that of women. This was thought to signify that men occupying a position "close to that of women" in the labor market tended to be unmarried. However, it is possible that this tendency disappears with an increase in age, and it is still unclear whether this will weaken occupational gender segregation. Japanese society exhibits trends of expanding non-regular employment and increasing unmarried population. Meanwhile, vertical occupational gender segregation among people with degrees remains high despite the expansion of higher education. Will this segregation change soon?

Improvements in the occupational status of women with human capital would reduce occupational gender segregation among highly educated individuals. The analysis in this chapter suggests that improvements in these areas require advances in post-industrialization and the elimination of women's disadvantages in the early career stages. The percentage of manual workers is higher, while the percentage of upper non-manual workers (managers, professionals, and technicians) is lower in Japan when these percentages are compared to other industrialized countries. Thus, the value of higher education in Japan is considered to be low. Furthermore, women are in occupations and employment status that are unrelated to social resources from the preliminary stages of their careers, partially because women's specializations in higher education differ from those of men. Moreover, further expansion of non-regular employment and service occupations may reduce occupational gender segregation to align the status of men with that of women. However, the importance of employment status as a form of occupational status in Japan and the large differences in social resources obtained by regular and non-regular employees have always been associated with gender policy. There is the possibility that the importance of employment status, which has been confirmed in this chapter, may change if the Japanese government and businesses truly pursue gender equality to change social policies and business practices.

Notes

1. However, the percentage of unmarried people has increased in recent years, and the number of women who do not experience interruptions in their careers is increasing.

2. "Professionals" and "technicians and associate professionals" cannot be distinguished in the major groups of the JSCO. As indicated in chapter "Multidimensional Structure of Occupational Segregation by Education and Gender: Time Series Comparison in Japan from 1980 to 2010", the distinction between professional occupations and technical occupations is important in describing occupational gender segregation among highly educated people in Japan.
3. The "self-employed/family employee" configuration is omitted.
4. Increasing opportunities for women to have professional and technical occupations generally signifies an increase in their occupational status. The odds ratios for male (vs. female) upper white-collar occupations (professional occupations and managerial occupations, excluding non-regular employees) by logistic regression analysis controlling for age group and educational level from 1992 to 2007 decreased from 1.79 in 1992 to 1.30 in 2007.
5. The "self-employed/family employee" configuration is omitted in Fig. 8.
6. Analyses using minor groups reveal that professional/technical occupations vary from those with many men to those with many women, which causes occupational segregation between highly educated men and women. If the data allow, it is necessary to divide professional occupations and technical (associate professional) occupations.
7. The 1997 and 2002 results are omitted.
8. Unmarried individuals in their 20s in 1992 and 1997 comprise a large fraction of married individuals in their 30s in 2002 and 2007, respectively.

Acknowledgements This research received grants from JSPS: 15K03818 and 18K01966. The anonymised data of the 1992, 1997, 2002, and 2007 Employment Status Surveys (Ministry of Internal Affairs and Communications) analysed in this chapter were provided by the National Center for Statistics, Japan. In addition, we wish to acknowledge the statistical office that provided the underlying data in this research.

References

Arita, Shin. 2016. Shotoku to shukanteki-chii-hyouka no kakusa (Disparity of income and subjective status evaluation: Are the effects of firm size and employment status strong?). In *Shugyo-kikai to houshu-kakusa no shakaigaku* (*Sociology of employment opportunity and income disparity*). Tokyo: University of Tokyo Press (in Japanese).

Bell, Daniel. 1973. The coming of post-industrial society. In *Social stratification*, 4th ed., ed. David B. Grusky, and Katherine R. Weisshaar, 2014, 1066–1077. Boulder: Westview Press.

Brinton, Mary C. 2001. Married women's labor in East Asian economies. In *Women's working lives in East Asia*, ed. Mary C. Brinton, pp. 1–37. Stanford: Stanford University Press.

Cabinet Office. 2019. White paper on gender equality 2019. https://www.gender.go.jp/english_contents/about_danjo/whitepaper/index.html.

Cejka, Marry Ann, and Alice H. Eagly. 1999. Gender-stereotypic images of occupations correspond to the sex segregation of employment. *Personality and Social Psychology Bulletin* 25 (4): 413–423.

Chang, Mariko Lin. 2000. The evolution of sex segregation regimes. *American Journal of Sociology* 105 (6): 1658–1701.

Charles, Maria, and David B. Grusky. 2004. *Occupational ghettos*. Stanford: Stanford University Press.

Esping-Andersen, Gøsta. 1999. *Social foundations of postindustrial economies*. Oxford: Oxford University Press.

Featherman, David L., Lancaster Jones, and Robert M. Hauser. 1975. Assumptions of social mobility research in the US: The case of occupational status. *Social Science Research* 4: 329–360.

Guinea-Martin, Daniel, Ricardo Mora, and Javier Ruiz-Castillo. 2018. The evolution of gender segregation over the life course. *American Sociological Review* 83 (5): 983–1019.

Hirasawa, Kazushi. 2011. Daigaku no gakkoureki wo kamishita kyouiku shokugyou tassei bunseki (The analysis of educational and occupational attainment focusing on educational credentials). In *Gendai no kaisou shakai 2: Kaisou to idou no kouzou* (*Stratification in contemporary society: The structure of stratification and mobility*), ed. Hiroshi Ishida, Hiroyuki Kondo, and Keiko Nakao, 155–170. Tokyo: University of Tokyo Press (in Japanese).

Hirata, Shuichi. 2011. Josei no life course to shugyo (Women's life course and work: the change of m-shape curve). *Gendai no kaisou shakai 2: Kaisou to idou no kouzou* (*Stratification in contemporary society: The structure of stratification and mobility*), ed. Hiroshi Ishida, Hiroyuki Kondo, and Keiko Nakao, 223–237. Tokyo: University of Tokyo Press (in Japanese).

Horowitz, Jonathan. 2018. Relative education and the advantage of a college degree. *American Sociological Review* 83 (4): 771–801.

Iwai, Hachiro. 2011. Gakureki to shoki career no doutai (Dynamics of education and early career: Changes in post-war Japanese life course). In *Gendai no kaisou shakai 2: Kaisou to idou no kouzou* (*Stratification in contemporary society: The structure of stratification and mobility*), ed. Hiroshi Ishida, Hiroyuki Kondo, and Keiko Nakao, 205–222. Tokyo: University of Tokyo Press (in Japanese).

Iwata, Masami, Machiko Osawa, and Nihon Joshi Daigaku Gendai Josei Career Kenkyujo (eds.). 2015. *Naze josei ha shigoto wo yamerunoka* (*Why do women quit their jobs?: Analysis of 5155 women's trajectory*). Tokyo: Seikyusha (in Japanese).

Kambayashi, Ryo. 2017. *Seiki no sekai, hiseiki no sekai* (*The world of regular employment, the world of non-regular employment*). Tokyo: Keiogijuku Daigaku Shuppankai (in Japanese).

Levanon, Asaf, and David B. Grusky. 2016. The persistence of extreme gender segregation in the twenty-first century. *American Journal of Sociology* 122 (2): 573–619.

Magnusson, Charlotta. 2009. Gender, occupational prestige, and wages: A test of devaluation theory. *European Sociological Review* 25 (1): 87–101.

Oda, Akeko, Reiko Yamato, and Hiroshi Taroumaru. 2014. Seibetsu-shokuiki-bunri ha chiiki niyotte dou kotonaruka (How do occupational gender segregations differ by region?). In *Higashi-Asia no roudou-shijo to shakai-kaisou* (*Labor market and social stratification in East Asia*), ed. Hiroshi Taroumaru, 137–168. Kyoto: Kyoto University Press (in Japanese).

Ohtsuki, Nami. 2015. *Shokumu kakusa* (*Job disparity: What factors hinder promotion of women's advancement?*). Tokyo: Keiso Shobo (in Japanese).

Osawa, Mari. 2007. *Gendai nihon no seikatsu hosho system* (*Life security system in contemporary Japan*). Tokyo: Iwanami Shoten (in Japanese).

Statistics Bureau of Japan. 2016. "Population census". https://www.stat.go.jp/english/data/kokusei/index.html.

Statistics Bureau of Japan. 2020. "Historical data (labour force survey)". https://www.stat.go.jp/english/data/roudou/lngindex.html.

Suzuki, Kyoko. 2018. Roudou-shijou no senzai-kouzou to koyou-keitai ga chingin ni ataeru eikyou (The latent structures of the labour market and the effect of employment categories on wages in Japan: A latent class analysis with finite mixture model). *Japanese Journal of Labour Studies* 698: 73–89 (in Japanese).

Taroumaru, Hiroshi. 2009. *Jakunen hiseiki-koyou no shakaigaku: Kaisou, gender, global-ka (Sociology of young non-regular workers: Stratification, gender, globalization).* Osaka: Osaka University Press (in Japanese).

Torche, Florencia. 2011. Is a college degree still the great equalizer? Intergenerational mobility across levels of schooling in the United States. *American Journal of Sociology* 117 (3): 763–807.

Treiman, Donald J. 1970. Industrialization and social stratification. *Sociological Inquiry* 40 (2): 207–234.

Uchikoshi, Fumiya, and Ryota Mugiyama. 2020. Nihon niokeru seibetsu-shokuiki-bunri no susei (Trends in occupational sex segregation in Japan: A decomposition analysis of census data, 1980–2005). *Journal of Population Studies* 56: 9–23 (in Japanese).

Yamaguchi, Kazuo. 2019. *Gender inequalities in the Japanese workplace and employment: Theories and empirical evidence.* Singapore: Springer.

Occupation Segregation among Immigrants in Japan: Using Custom-Made Census Data in 2010

Shigemi Ohtsuki

1 Accepting Immigrants

As summarized in the chapter "Occupational Segregation by Education and Gender in Japan: Focusing on Employment Status and Marital Status" of this book, Charles and Grusky (2004) and Oda et al. (2014) have pointed to occupational segregation by gender in postindustrial societies. Their paradox that attributes become more salient in a society with meritocracy has been confirmed in the previous chapters of this book. The change in the relationship between attributes and achievements through industrial transformation requires us to reconsider this from a different perspective: a reconsideration of the connection between attributes excluding gender, such as nationality, ethnicity, and occupational performance.

This chapter aims to capture the correspondence between the educational background and occupation of immigrants in Japanese society. Despite losing World War II, Japan became one of the largest economies in the world, partly due to external factors such as the Korean War. Sassen (1991) cited Tokyo as a global city, intimating that Japanese society may be positioned as the earliest successful postwar Asian society on the trajectory of modernization.

Diversity is a cornerstone for development in Western societies. Ethnic diversity has not been denied, although a retreat from multiculturalism has been pointed out (Banting and Kymlicka 2013). On the other hand, one might say that homogeneity has been a linchpin for development in Japan. In fact, Japan is said to be a society with

Note that the Japanese census does not include in its data members of diplomatic missions and consular agencies of foreign governments and their families, and military personnel and their families in foreign armed forces. Therefore, they are excluded from the analysis.

S. Ohtsuki (✉)
Department of Urban Science and Policy, Faculty of Urban Environmental Sciences, Tokyo Metropolitan University, 1-1 Minamiosawa, Hachioji City, Tokyo 1920397, Japan
e-mail: ohtsuki@tmu.ac.jp

K. Nakao (ed.), *Multidimensional Aspects of Occupational Segregation*, Behaviormetrics: Quantitative Approaches to Human Behavior 18,
https://doi.org/10.1007/978-981-99-8513-5_4

high ethnic homogeneity compared to other countries (Alesina et al. 2003; Fisher 2013).

Against this backdrop, Japan has finally begun accepting immigrants in earnest in recent years to supplement its workforce in the face of a declining population. Specifically, the Japanese government began accepting immigrants as unskilled workers in 2018 by expanding the range of residence status, and the Japanese government announced that it would accept more than 500,000 immigrants by approximately 2025 (Yamada 2018). In addition to unskilled workers, the government is actively accepting highly skilled human resources and international students who can be regarded as their protégés.[1]

Accepting immigrants in Japan is not a recent phenomenon. Since their defeat in World War II, the Japanese population has become more educated, and it has become difficult to secure unskilled workers from the domestic population for the manufacturing industry. At the same time, in response to the declining birthrate and aging population, the Japanese government has selectively opened the doors to immigrants since the 1990s by accepting the Japanese descendants and utilizing the technical internship system. Therefore, one can conclude that there is a certain benefit in accepting immigrants in postwar Japan. Nevertheless, the official announcement of the acceptance of unskilled immigrants represents a significant shift in Japanese society, which has previously been highly ethnically homogeneous.

However, it is still unclear whether immigrants will be incorporated into Japanese society. The Japanese government does not use the term "immigrant" but "foreign nationals" in official situations. This can be interpreted to mean that, except for highly skilled personnel, people who come to Japan from abroad are expected to stay for a few years and then return to their home country or move to another country. In fact, in 2018, Prime Minister Shinzo Abe made the statement, "We are not thinking of maintaining the nation by accepting foreigners and their families of a certain size relative to the national population without setting a time limit" (Asahi Newspaper 2018).

Can Japan become an ethnically diverse society or will immigrants stay only temporarily, as the government designs? Moreover, will the ethnic homogeneity of the society be maintained? It is necessary to understand the current situation of immigrants from the perspective of social stratification to understand whether Japanese society can promote social integration that incorporates immigrants. In this chapter, the differences in social stratification between Japanese and immigrants are examined by an exploratory investigation of the relationship between education and occupation.

2 Migration Studies in Japanese Sociology

In this section, relying on Ohtsuki (2022), I summarize the trends and issues in studying migration in Japanese sociology.[2] Japanese society has consistently suffered from labor shortages since its defeat in World War II. From the post-war reconstruction to the 1960s, a period of rapid economic growth, the demand for labor was met by a demographic shift from rural to urban areas. However, as postwar Japan became more affluent and the number of well-educated Japanese increased, it became difficult to acquire blue-collar workers such as those in the manufacturing industry. Against this backdrop, immigration gradually increased in the 1980s.

With this background, Japanese sociology has paid attention to immigration since the late 1980s. Supposing that ethnicity, an attribute of a highly industrialized modern society, attracted attention in American sociology in the 1950s (e.g., Bogardus 1958; Gordon 1964; van den Berghe 1976), Japanese sociologists' attention to immigration was somewhat delayed, with a few exceptions. This delay itself suggests the ethnically homogeneous aspect of Japanese society.

This section categorizes migration studies into policy approaches to immigrant incorporation, local community approaches, and the labor market and social stratification approach. Undeniably, these are not mutually exclusive, but complementary and overlapping.

Policy approaches view citizenship, welfare, and multiculturalism policies in the process of integrating immigrants into host societies (e.g., Banting and Kymlicka 2013; Goodman 2010). In Europe, institutionalizing the possibility of social integration has been recognized as an urgent issue. Recent immigration studies in Japan have actively examined the incorporation of immigrants from a policy approach (e.g., Higuchi 2016; Hollifield and Sharpe 2017; Milly 2014; Nagayoshi 2020; Oishi 2018; Takahata 2016) . In Japan, the focus is on acceptance policies, including language support and systems such as pensions, employment insurance, and other social security. Such an approach points out Japan's immigration reception inadequacies in terms of institutions, as Hollifield and Sharpe (2017) describe Japan as economically open but politically and rights-closed.

The local community approach discusses integration beyond individual nationalities and ethnic groups in regions and daily life. In Japan, this focus on local communities has been actively promoted by urban and regional sociology (e.g., Okuda and Tajima 1991; Liu-Farrer 2011; Shipper 2008).[3] An academic interest in population heterogeneity, from the perspective of Japanese society, and a practical expectation of immigrants as new members of local communities in the face of the aging population and an increasing need for welfare originated in the Chicago School. Shipper conducted a qualitative study focusing on the adjustment of several immigrant groups and proposed a hierarchical structure of immigrant groups. According to Shipper (2008), Korean and Japanese Americans from South America, followed by immigrants from China and Southeast Asia, are privileged in Japanese society.

However, Shipper's work was published in 2008. Given the changes in the global economy due to the bankruptcy in 2008 and changes in residency status, the pecking order of immigrant groups presented by Shipper (2008) may have changed. Especially in the 2010s, the number of Vietnamese people increased, driven by expanding residency status. In addition, Filipinos, who previously came to Japan as entertainers, are increasingly arriving as nurses and caregivers against the Japan-Philippines Economic Partnership Agreement. Therefore, it is necessary to confirm the hierarchical structure of immigrant groups, considering the current situation in Japanese society.

It can be said that Japanese migration studies with policy and local community approaches have been conducted in the same way as those in Europe and the U.S. However, the labor market and social stratification approaches have been stagnant. This stagnation is not a theoretical problem but a problem about data. In fact, in the early stages of Japanese immigration studies, Kajita (1988) organized the concept of ethnicity according to the relationship between achieved status and ascribed status. He argued that ethnicity, as an ascribed status, did not disappear but became newly problematic and active in modernization, i.e., the selection of meritocracy as the main allocation principle. Therefore, sociologists in Japan adopted the theoretical framework of the social stratification approach in the initial stages of Japanese immigration studies.

However, subsequent Japanese immigration studies have focused on the policy and local community approaches. There has been a reluctance to adopt a labor market or social stratification approach.[4] This difference can be considered a significant distinction between European and American migration studies. For example, European and American studies such as Portes (1993, 2001) examine the adaptation of the second generation of immigrants in American society from the perspective of segmented assimilation and point out that adaptation processes differ among immigrant groups. In Europe, the educational attainment of immigrants and their status in the labor market have been examined through comparisons among European countries using systematic data such as the Organisation for Economic Co-operation and Development (OECD) Programme for International Student Assessment (PISA) (Heath et al. 2008).

The stagnation of the labor market and social stratification approaches in Japanese migration studies is not because the theoretical frameworks adopted in Western migration studies are not suitable for Japan's migrant society. Instead, it is simply a matter of data. Mainstream Japanese immigration research is either qualitative research or examinations of policies and institutions and has not yet reached the stage of conducting quantitative social surveys and statistically examining the positions of immigrants in social stratification. Obtaining a sampling list with the addresses of immigrants has been challenging. Consequently, it has been impossible to conduct a quantitative social survey at the national level based on random sampling. However, since 2009, the census has been available for tailor-made tabulation. Although there are data limitations, it has become relatively easy to analyze the available data using the labor market and social stratification approaches.

Ohtsuki (2022) explores the relationship between education and occupation between Japanese and Chinese immigrants through a correspondence analysis using census data. The results of the analysis show that Chinese immigrants effectively use their educational background in response to the demand for information technology and other professional occupations. The results further suggest that meritocracy, such as the stronger association between education and occupation, is partially valid in the case of immigrants. Education is found to be a valuable resource for immigrants who face hurdles in performing vertical social mobility in the host society, including issues such as language and cultural values. However, Ohtsuki's analysis only compared Japanese and Chinese immigrants. It does not clarify the position of immigrant groups in Japanese society, and it is necessary to compare multiple immigrant groups simultaneously. Therefore, in this chapter, I will focus on the hierarchical status of immigrants and examine the hierarchical structure of current Japanese society through quantitative data analysis. Specifically, I extend Ohtsuki's (2022) research by conducting a correspondence analysis between education and occupation for multiple immigrant groups.

3 Framework, Methods, and Variables

This chapter focuses on immigrants in Japanese society and examines the hierarchical structure of Japan from the perspective of the Japanese and their hierarchical status. I capture the correspondence between educational background and occupation. Specifically, the analysis adopts the following steps: First, correspondence analysis between education and occupation is conducted using the integrated data of Japanese and all immigrants mentioned above. Subsequently, individual data for each immigrant group are plotted as supplementary points. This chapter only presents the supplementary points for each immigrant group due to space limitations. Gender, age, education, and nationality are the attributes in the correspondence analysis. The age categories were 20–34 years old or younger, 35–49 years old, and 50–64 years old. There are four sub-categories for education: elementary/middle school, high school, junior college/technical college, and university/graduate school. This means that there are 24 attributes overall. On the other hand, the occupation category is based on the classification in Hosokawa(2018). Hosokawa reclassified occupational categories in the census into 68. Since Hosokawa's classification uses occupational classification in the census, it is considered to reflect the actual occupational structure in Japan. At the same time, Hosokawa's classification is designed to be converted to the International Standard Classification of Occupations. Our analysis focuses on immigrant groups from the following locations:Korea, China, the Philippines, Vietnam, the U.S., and Brazil. The census also collects data on several other immigrants' nationalities. However, based on the findings of Shipper (2008), we limit our analysis to the above groups. In addition to the above groups of immigrants, Japanese people are included in the analysis.

4 Data

In this chapter, we use customized aggregated census data. It is possible to access aggregate census data from the Statistics Bureau of the Ministry of Internal Affairs and Communications. However, only a few statistical tables are available. In addition, researchers can view only the aggregate data for each variable; it is not easy to analyze the relationship between variables on an individual basis. In pursuing this chapter's research question, more is needed to analyze the relationship between education and immigrants' occupations only from the statistical tables provided. Therefore, we use the custom-made service of the census.

The law about the availability of census information was amended in 2009 in Japan. It made it possible to obtain publicly available data and custom-made tabulated data if it is deemed to contribute to academic research and higher education development. This custom-made tabulation service allows for analyzing the relationship between education and occupation for each immigrant group. We used data from the 2010 census.

Individual cases should be mentioned here. Special permanent residents are people from the Korean Peninsula and their descendants and Chinese from China and Taiwan and their descendants who have lived in Japan before World War II. Therefore, Chinese and Korean immigrants are a mixture of oldcomers, including special permanent residents and newcomers who came to Japan to work or study in recent years. Special permanent residents, like Japan's permanent residents, have no employment restrictions. According to the national census, about 99% of the special permanent residents are immigrants of Korean descent.

In addition, there are many cases in which Brazilians come to Japan as long-term residents. Strictly speaking, if a person's ancestors are Japanese and of Japanese descent, they can obtain long-term resident status. Specifically, Japanese-Brazilians, Japanese-Peruvians, and Japanese-Filipinos fall under this category. In the case of Brazilians working in Japan, many can be considered Japanese-Brazilians. Long-term residents are qualified to stay based on status or position, not their activities. Therefore, like special permanent residents, they have no restrictions for employment. Hence, the employment characteristics of Korean and Japanese-Brazilian immigrants differ from those of other immigrant groups, as they consist of several special permanent residents and long-term residents. In other words, as Shipper (2008) indicated, these two groups are in a better position to engage in socioeconomic activities than other immigrant groups.

5 Analysis

Figure 1 shows the distribution of occupations in the correspondence analysis. The contribution rate in the first dimension is 40.3% and is 26.9% in the second dimension. Each letter of the alphabet indicates the major group level of the occupation, i.e., A

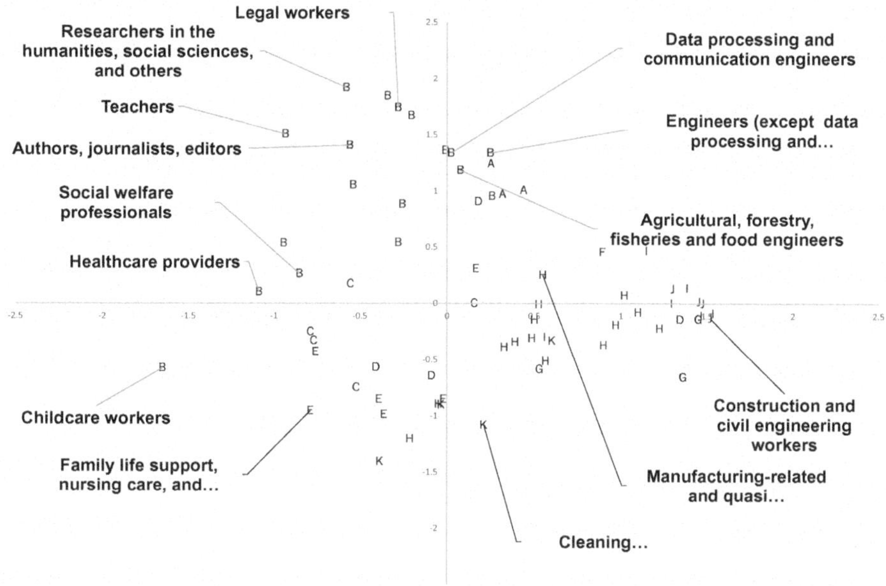

Fig. 1 Configuration of Fig. 1 Configuration of occupational status in 2010. Note: The first dimension's contribution is 0.403, and the second dimension's contribution is 0.269

management, B professional and technical, C clerical, D sales, E service, F security, G agriculture, forestry, and fishery, H production process, I transportation and machine operation, J construction and mining, and K transportation, cleaning, packaging, etc.

Professional jobs are plotted in the second quadrant, and managerial jobs are plotted near the vertical axis in the first quadrant. On the other hand, blue-collar jobs are plotted in the third or fourth quadrant, including production processes in the fourth quadrant. Therefore, broadly speaking, we can discern that occupational status is plotted along a negative slope.

Furthermore, we can roughly identify two parallel lines, upper and lower, for the negative slope plots from the second to the fourth quadrant. In the lower part, we can distinguish where those so-called "feminine" occupations are, as professional and technical occupations such as social welfare professionals and healthcare workers are plotted. In contrast, typical "masculine" occupations, such as production-related, are plotted in the upper row. Thus, the occupational plots in Fig. 1 can be interpreted as the axis of occupational status and gender track. That is, we can conclude that the occupational distribution from the upper left quadrant to the lower right quadrant is characterized by occupational status and gender track. Based on these characteristics, we now consider the distribution of attributes by nationality.

Figures 2, 3, 4, 5, 6, and 7 show the distribution of education by nationality. The correspondence analysis results, including all nationalities, are indicated, and the data for each nationality are output as a supplementary category. Because of space limitations, the output that includes all nationalities is omitted. However, based on the population ratio, the whole data is expected to approximate the plot for the Japanese. Therefore, when we plot the distribution of education by nationality, we plot Japanese education as a comparison. Letters represent nationalities: Japanese (J), Korean (K), Chinese (C), Filipino (P), Vietnamese (V), American (U), and Brazilian (B). As a comparison, Japanese (J) had their colors reversed. For all nationalities, the letter is enclosed in squares for males and left unchanged for females.

If we take a broad view of the distributions in Figs. 2, 3, 4, 5, 6 and 7, we can see that the correspondence between education and occupation differs for each immigrant group. The distribution for Koreans resembles that of the Japanese (Fig. 3). In other words, in the case of Koreans, a high level of education is more likely to lead to high-status occupations such as professional positions. The link between education and occupation is not as large for the Japanese as it is for other immigrant groups. Therefore, it can be said that the Korean educational background is valued in almost the same way as that of the Japanese when it comes to obtaining a job. This similarity was also ascertained for gender segregation. In other words, in the case of Koreans, no matter how educated they are, there is a clear division in Japanese society regarding the connection between education and occupation, just as there is with the Japanese.

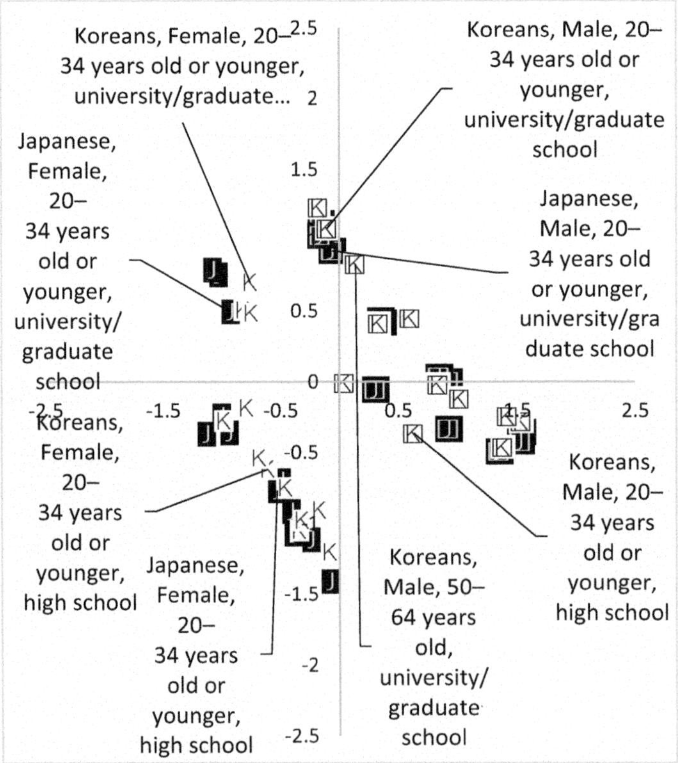

Fig. 2 Configuration of education of the Koreans

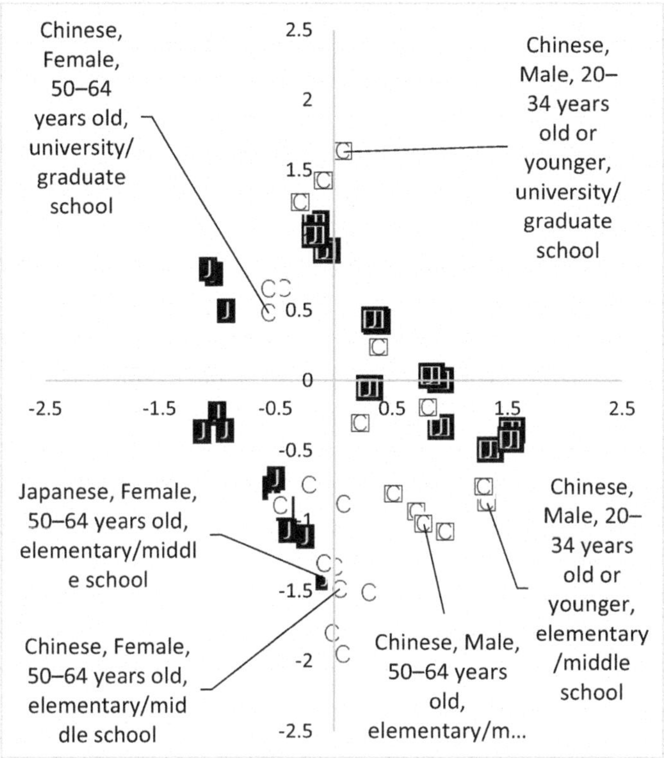

Fig. 3 Configuration of education of the Chinese

The Chinese, too, have a similar plot to the Japanese, though not as close in resemblance as the Koreans. .Highly educated males are plotted in a location similar to highly educated Japanese males. However, in the case of the Chinese, unlike the Koreans, the plot of highly educated women is relatively close to that of men. Even in the case of the less educated, it can be concluded that the plots of Chinese men and women are relatively closer to each other than those of Japanese men and women. Hence, we can say that the division by gender is relatively small among the Chinese.

In the case of the Americans, as in the Chinese, we can say that the division by gender is small, and this is especially true for the highly educated. In fact, in the case of the highly educated, both males and females are plotted in similar places, indicating that higher education is associated with certain professions. Based on the plots of occupations in Fig. 1, we can assume that the plots of highly educated people correspond to that of teachers. In other words, professional life as an English teacher is typical of highly educated Americans adapting to Japanese society. In addition, the plots of Americans, both male and female, resemble those of highly educated Japanese, even if they have a moderate level of education. We can interpret that the

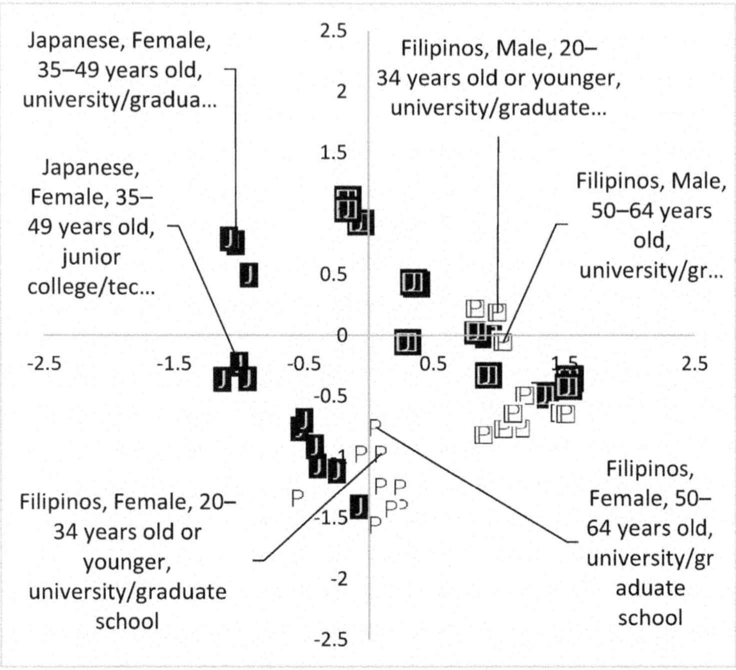

Fig. 4 Configuration of education of the Filipinos

advantage of English is meaningful enough in Japanese society to compensate for the educational background.

The Vietnamese also show a correspondence between higher education and professional occupation in the case of men. Regarding women, it can be pointed out that education and occupation correspond differently from Japanese women. However, because young, highly educated people are plotted near the origin, I would like to refrain from making a positive interpretation of this correspondence analysis. At least in Vietnamese, higher education is associated with higher occupation, only for males. However, since the plot of the highly educated young male population is in the first quadrant, it is necessary to examine its detailed correspondence with occupations. This is discussed later in Tables 1 and 2.

In the above immigrant groups, higher education was associated with professional occupations. Although the range of occupations they could choose may be narrow, we can conclude that the high-education immigrant group was associated with high-status occupations. However, other groups of immigrants live differently in Japan, Filipinos, and Brazilians.

The highly educated Filipinos and Brazilians are distributed around the fourth quadrant, indicating that the correspondence between education and occupation differs from the immigrant group described above. In other words, in the case of

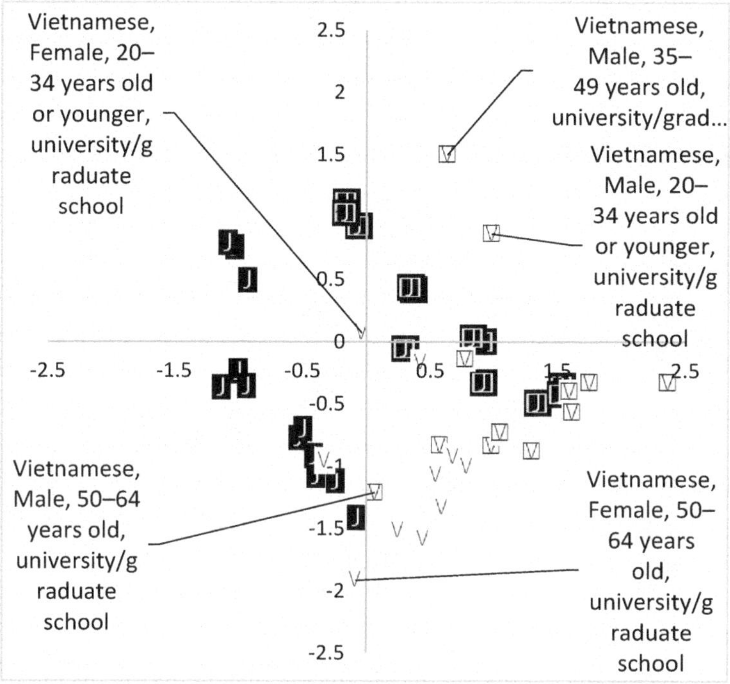

Fig. 5 Configuration of education of the Vietnamese

this immigrant group, high education does not correspond to high-status occupations. Thus, it can be interpreted that the meritocracy does not function. In the case of Brazilians, although their status as long-term residents allows them to engage in free economic activities, their educational background does not function as a route to high occupational status.

The analysis results in Figures 1 through 7 can be summarized as follows: First, there are two groups of immigrants: those whose education and occupation correspond to each other and those whose education and occupation do not. Immigrants from East Asia resemble the Japanese in that their educational background corresponds to their occupation. On the other hand, Filipinos and Brazilians are immigrant groups whose education does not correspond to their occupation, and meritocracy does not function. In the case of Vietnamese men, who have immigrated in increasing numbers in recent years, education is linked to occupation, and meritocracy functions partially. With Americans, there is a tendency to associate higher status with the occupation than with the individual's education. In addition, except for Americans, gender segregation of occupations exists among many immigrant groups, even when they are highly educated.

From the above analysis, several patterns emerged in which higher education corresponded to professional positions. Knowing the specific professions associated

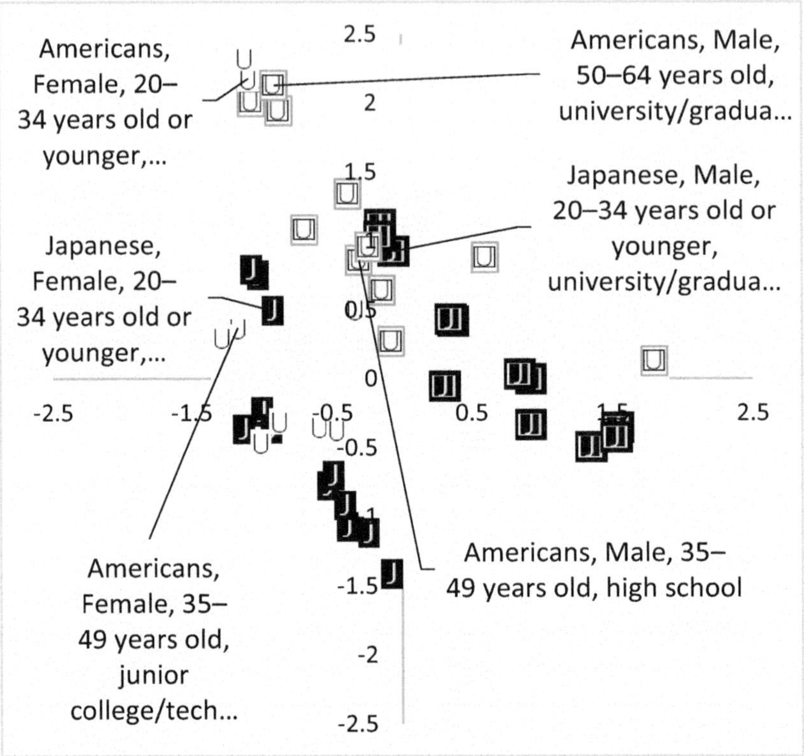

Fig. 6 Configuration of education of the Americans

with them is necessary to deepen the interpretation. Therefore, Tables 1 and 2 illustrate the top five occupations with the highest percentage of employment among the younger and older age groups, focusing on the highly educated in each immigrant group. Occupations classified as professionals are highlighted in inverted colors and fonts. The last row shows the total percentage of professionals in the top five occupations.

Let us begin with men (Table 1). Among Japanese men, engineers and information technology (IT) engineers are typical professions in which young people are engaged. Teachers are listed among the older generation, but not among the top five occupations among the younger generation, suggesting a response to changes in labor demand due to macroeconomic structural changes, such as the rapid growth of the IT industry. Among the immigrant group of highly educated males, engineers and IT engineers are among the top five Korean, Chinese, and Vietnamese occupations. The percentage of IT engineers is higher than that of Japanese engineers, especially in the case of the Chinese. In other words, they can be described as the group that best responds to the rapid progress of the IT industry. While some Vietnamese are engaged in IT engineering, others are engaged in manufacturing. In other words, there is a

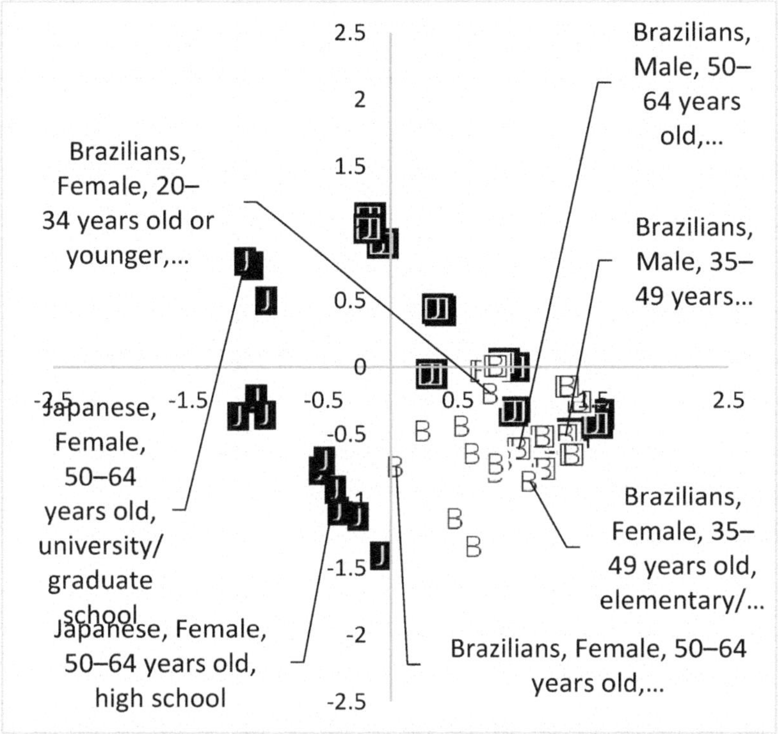

Fig. 7 Configuration of education of the Brazilians

mix of white-collar and blue-collar workers among the highly educated. We can say that highly educated Vietnamese men were plotted in the first quadrant in the correspondence analysis above because some highly educated people were engaged in such blue-collar jobs.

In contrast, IT engineers are not listed in the top five occupations for Filipinos and Brazilians, and the percentage of total professionals is not as high as that of Japanese. In other words, they have not responded to the demand for specialists that has accompanied the shift in the industrial structure, and even highly educated people are being forced to fill the demand for blue-collar jobs.

A high percentage of Americans, both young and old, are teachers or private tutors, indicating that they are consistently engaged in occupations that use their English language skills. However, among the younger groups, other professions have overtaken teaching as the number one occupation. This change indicates that highly educated Americans may be beginning to respond to the demand for professionals other than teachers that has arisen with the diversification of the industrial structure, such as the development of the service industry.

Now, we examine the results for women (Table 2). For well-educated Japanese women, "feminine occupations," such as healthcare providers and teachers, are listed

Table 1 Top five occupations of highly educated males

Japanese, Male			
20–34 years old (Total N = 2612480)		50-64 years old (Total N = 3031730)	
Outdoor service workers	16.8%	Clerks	21.9%
Clerks	16.1%	Outdoor service workers	12.5%
Engineers (except Data processing and communication engineers)	**8.2%**	**Teachers**	**8.3%**
Data processing and communication engineers	**7.9%**	Executive officers of corporations and associations	7.7%
Merchandisers	5.6%	**Engineers (except Data processing and communication engineers)**	**5.4%**
Total professional and technical occupations	**16.2%**	**Total professional and technical occupations**	**13.8%**
Korean, Male			
20–34 years old (Total N = 7780)		50-64 years old (Total N = 6600)	
Outdoor service workers	17.5%	Executive officers of corporations and associations	19.1%
Clerks	14.3%	Outdoor service workers	10.6%
Data processing and communication engineers	**12.1%**	**Healthcare providers**	**9.8%**
Engineers (except Data processing and communication engineers)	**8.0%**	Clerks	9.4%
Food, beverage and other service workers	5.8%	Vehicle drivers	4.5%
Total professional and technical occupations	**20.1%**	**Total professional and technical occupations**	**9.8%**
Chinese, Male			
20–34 years old (Total N = 14970)		50-64 years old (Total N = 2920)	
Data processing and communication engineers	**31.8%**	Outdoor service workers	12.7%
Engineers (except Data processing and communication engineers)	**14.7%**	**Teachers**	**12.3%**
Clerks	10.7%	Executive officers of corporations and associations	9.6%
Outdoor service workers	9.0%	**Healthcare providers**	**9.6%**
Other professional workers	**4.0%**	Clerks	9.2%
Total professional and technical occupations	**50.5%**	**Total professional and technical occupations**	**21.9%**
Phillippines, Male			
20–34 years old (Total N = 2330)		50-64 years old (Total N = 390)	
Metal and other product manufacturing and processing workers	20.6%	Construction and civil engineering workers	12.8%
Engineers (except Data processing and communication engineers)	**11.6%**	**Engineers (except Data processing and communication engineers)**	**10.3%**
Transport machinery assembly and repair workers	8.6%	Metal and other product manufacturing and processing workers	10.3%
Electro-mechanical apparatus assembly maintenance and repair workers	8.2%	Food, beverage and cigarette manufacturing workers	7.7%
Rubber, plastic product manufacturing workers	5.6%	Vehicle drivers	7.7%
Total professional and technical occupations	**12%**	**Total professional and technical occupations**	**10%**
Vietnamese, Male			
20–34 years old (Total N = 1130)		50-64 years old (Total N = 60)	
Engineers (except Data processing and communication engineers)	**20.4%**	Food and drink preparatory workers	66.7%
Manufacturing-related and quasi manufacturing workers	12.4%	Transport and postal clerks	33.3%
Data processing and communication engineers	**11.5%**		0.0%
Metal and other product manufacturing and processing workers	10.6%		0.0%
General machinery and apparatus assembly and repair workers	10.6%		0.0%
Total professional and technical occupations	**31.9%**	**Total professional and technical occupations**	**0%**
American, Male			
20–34 years old (Total N = 3500)		50-64 years old (Total N = 2290)	
Other professional workers	**34.0%**	**Teachers**	**31.4%**
Teachers	**20.6%**	**Private tutors**	**14.8%**
Private tutors	**20.3%**	Clerks	12.2%
Clerks	8.6%	**Religious services**	**6.1%**
Outdoor service workers	4.3%	Executive officers of corporations and associations	5.2%
Total professional and technical occupations	**74.9%**	**Total professional and technical occupations**	**52.4%**
Brazilian, Male			
20–34 years old (Total N = 570)		50-64 years old (Total N = 1240)	
Electro-mechanical apparatus assembly maintenance and repair workers	17.5%	Food, beverage and cigarette manufacturing workers	17.7%
Transport machinery assembly and repair workers	14.0%	Metal and other product manufacturing and processing workers	13.7%
Other professional workers	**8.8%**	Transport machinery assembly and repair workers	10.5%
Metal and other product manufacturing and processing workers	8.8%	Carrying workers	5.6%
Clerks	7.0%	Rubber, plastic product manufacturing workers	4.8%
Total professional and technical occupations	**8.8%**	**Total professional and technical occupations**	**0%**

as professional occupations. Engineers and IT engineers, which are found in the case of men, are not found for women, even in the younger groups. Teachers and healthcare providers were also listed for Koreans, who showed a similar distribution to that of the Japanese in the correspondence analysis described above. IT engineers also appear in the list for the Chinese and Vietnamese. Therefore, some immigrant groups respond more successfully than the Japanese to the demand for high-level human resources

Table 2 Top five occupations of highly educated females

Japanese, Female

20–34 years old (Total N = 1681270)		50-64 years old (Total N = 668760)	
Clerks	33.7%	**Teachers**	**22.0%**
Healthcare providers	**10.8%**	Clerks	20.8%
Teachers	**8.3%**	**Healthcare providers**	**10.8%**
Merchandisers	7.2%	**Private tutors**	**5.1%**
Outdoor service workers	6.2%	Merchandisers	5.1%
Total professional and technical occupations	**19.1%**	**Total professional and technical occupations**	**37.9%**

Korean, Female

20–34 years old (Total N = 5310)		50-64 years old (Total N = 2100)	
Clerks	30.5%	Clerks	14.3%
Teachers	**7.5%**	**Teachers**	**12.4%**
Merchandisers	7.2%	Executive officers of corporations and associations	10.0%
Healthcare providers	**7.0%**	**Healthcare providers**	**9.0%**
Data processing and communication engineers	**6.4%**	Food and drink preparatory workers	6.7%
Total professional and technical occupations	**20.9%**	**Total professional and technical occupations**	**21.4%**

Chinese, Female

20–34 years old (Total N = 12290)		50-64 years old (Total N = 1310)	
Clerks	29.4%	Teachers	16.8%
Data processing and communication engineers	**12.0%**	Clerks	13.7%
Outdoor service workers	7.3%	Merchandisers	9.2%
Merchandisers	7.1%	Executive officers of corporations and associations	6.1%
Other professional workers	**6.1%**	Food, beverage and other service workers	5.3%
Total professional and technical occupations	**18.1%**	**Total professional and technical occupations**	**0%**

Phillippines, Female

20–34 years old (Total N = 2830)		50-64 years old (Total N = 1170)	
Food, beverage and cigarette manufacturing workers	12.7%	Food, beverage and cigarette manufacturing workers	20.5%
Food, beverage and other service workers	11.7%	Electro-mechanical apparatus assembly maintenance and repair workers	6.0%
Electro-mechanical apparatus assembly maintenance and repair workers	8.8%	Clerks	6.0%
Family life support and nursing care service	8.5%	Food, beverage and other service workers	5.1%
Rubber, plastic product manufacturing workers	6.7%	Metal and other product manufacturing and processing workers	5.1%
Total professional and technical occupations	**0%**	**Total professional and technical occupations**	**0%**

Vietnamese, Female

20–34 years old (Total N = 430)		50-64 years old (Total N = 20)	
Clerks	23.3%	Food, beverage and cigarette manufacturing workers	100.0%
Other professional workers	**9.3%**		0.0%
Other product manufacturing workers	9.3%		0.0%
Personal grooming and hygiene services	9.3%		0.0%
Data processing and communication engineers	**7.0%**		0.0%
Total professional and technical occupations	**16.3%**	**Total professional and technical occupations**	**0%**

American, Female

20–34 years old (Total N = 1660)		50-64 years old (Total N = 530)	
Other professional workers	**55.4%**	**Teachers**	**47.2%**
Teachers	**18.7%**	**Private tutors**	**13.2%**
Private tutors	**14.5%**	**Other professional workers**	**9.4%**
Clerks	3.0%	**Religionists**	**7.5%**
Outdoor service workers	2.4%	**Authors, journalists, editors**	**7.5%**
Total professional and technical occupations	**88.6%**	**Total professional and technical occupations**	**84.9%**

Brazilian, Female

20–34 years old (Total N = 460)		50-64 years old (Total N = 700)	
Electro-mechanical apparatus assembly maintenance and repair workers	21.7%	Food, beverage and cigarette manufacturing workers	34.3%
Transport machinery assembly and repair workers	13.0%	**Other professional workers**	**15.7%**
General machinery and apparatus assembly and repair workers	10.9%	Electro-mechanical apparatus assembly maintenance and repair workers	7.1%
Food, beverage and cigarette manufacturing workers	8.7%	Metal and other product manufacturing and processing workers	5.7%
Metal and other product manufacturing and processing workers	8.7%	Carrying workers	5.7%
Total professional and technical occupations	**0%**	**Total professional and technical occupations**	**15.7%**

in the IT industry in Japanese society. For women, traditional gender norms persist in Japanese society, and previous studies have pointed out that this makes it difficult for immigrant women to operate in the Japanese labor market (Liu-Farrer 2011). Considering this, the demand for professional positions in the IT industry may be becoming favorable for well-educated immigrant women.

However, not all immigrant groups are engaged in such specific professions. Typical blue-collar jobs such as manufacturing and assembly are listed among the top five occupations for well-educated Filipinas and Brazilian women. On the other hand, even among younger groups, there are no professional occupations such as engineers or IT engineers. Among Filipinas, nursing care workers are listed as the fourth most important occupation among young people, indicating a tendency to respond to the demand for labor in the silver industry. However, in the Japanese occupational classification, nursing care workers are classified as service workers rather than as professionals. Therefore, that the educational background of the Filipinas, like the that of Brazilians, is not being properly valued.

For educated American women, the list of occupations approximates that of educated American men. This approximation is consistent with American highly educated women being plotted close to highly educated men in the correspondence analysis. In other words, other professionals, teachers, and private tutors are listed, and it can be concluded that engaging in professions that make use of English is also practiced by women.

6 Conclusion

This chapter focuses on immigrants in Japanese society and captures the correspondence between the Japanese and their education and occupation. The 2010 census data were used in the analysis. Based on this analysis, the following points became clear. First, the correspondence between education and occupation differed depending on the immigrants' country of origin. Second, there was a notable correspondence between professional occupations, such as information processing technology and higher education, in the case of immigrants. Third, in the case of the Japanese, there was a clear gender track, but in some immigrants, such a gender track was unclear.

Some immigrants, including highly educated Chinese, tend to work as IT engineers and other professional occupations. Their choice implies that in Japan's social structure, where education, gender, and ethnicity are multilayered, education functions as a means for immigrants to obtain high-status jobs, even if partially. However, higher education was not associated with higher status occupations for some groups, such as Brazilian immigrants. In other words, it does not work equally among all ethnic groups. Thus, the function of education as meritocracy operates partially through ethnicity.

This point is also related to the segmented assimilation theory (Portes 1993, 2001). The segmented assimilation theory emphasizes that the adaptation of the second generation in the host society is not uniform, but the process is segmented for each ethnic group. The hierarchical situation of immigrants in Japanese society shows the same tendency. At the same time, this implies the difficulty of building a consensus for social integration. The Japanese government has stated that it aims to achieve equality between Japanese and foreign nationals, regardless of ethnicity. However, suppose the gap between such ideals and reality differs from the immigrant-to-immigrant group. In that case, it is not easy to establish a claim for the rights of an entire immigrant group.

If we share that the correspondence between education and occupation differs among immigrant groups in Japanese society, there is an opportunity to improve the system. The link between education and occupation differs according to ethnicity and nationality conflicts with modern values. Therefore, such pre-modern habits and institutions can be objects to be overcome by those who accept modern values. Particularly in the twenty-first century, Japanese society has entered the universal stage of education, with the percentage of students entering higher education institutions exceeding 50%. It is necessary to continuously observe whether a consensus can be established to improve the socio-economic conditions of immigrants through a high standard of education based on modern values.

In addition to promoting the improvement of the socio-economic conditions of immigrants from the host society through education, as mentioned above, the socio-economic activities of immigrants may promote the transformation of values in the host society. As noted in previous studies, traditional gender norms persist in the Japanese labor market; hence, immigrants who grew up in a society that places value on gender equality may find it difficult to accept working conditions in Japanese companies with a clear gender divide (Liu-Farrer 2011). For such well-educated immigrant women, actively developing their careers as professionals is consistent with their values. At the same time, many highly educated Japanese women may not welcome the gender track as much as immigrant women. Therefore, if immigrant women who have attained the same level of education as themselves achieve higher professional status in Japanese society, the activities of such immigrant women can serve as role models for Japanese women. Specifically, the success of immigrant women may provide an opportunity for Japanese society to reconsider the expression of "gendered selves" (Charles and Bradley 2009). Thus, immigrants are no longer mere helpers to supplement thc labor force; they are also a guide to ensure people realize the flawed nature of meritocracy in Japanese society.

However, it should be argued that the success of immigrants in Japanese society is limited in two ways. The first relates to the choice of occupation, and the second is the immigrant's nationality. The second point can be read as a restriction by the state as it relates to the status of residence. Regarding the choice of occupation, the occupation of successful immigrants is limited to IT engineers and other professional occupations. Some immigrants are provided with a route to a higher professional status if they can obtain a high level of education. However, professional and white-collar jobs, such as management positions, are an option for a Japanese male with the same education. In the case of immigrants, the number of such choices differs from that available to the Japanese. In other words, the value of freedom of choice in modern society is undermined.

The analysis results in this chapter show that the nationality of immigrants has an important meaning in obtaining occupational status in Japanese society. The reason nationality is so important is that immigrants' status of residence is linked to their occupation. Currently, there are 29 types of residence status in Japan (Immigration Services Agency 2023). Resident status can be divided into two main categories:

activity-based resident status and status-based resident status. Status-based resident status has no restrictions on socioeconomic activities, except certain types of work, such as public service. Thus, freedom of occupational choice is ensured. However, the activity-based resident status limits the possibility of changing jobs, as the content of your work is determined in advance.[5] If immigrants have permanent resident status, they will have no restrictions on changing their jobs. However, immigrants usually need to stay in Japan for ten years to obtain permanent resident status, which is a lengthy period when considering one's career plan. In other words, it is more attractive for immigrants to move to a society that appreciates them more or to return to their home country than to continue living in Japan (Liu-Farrer 2011). Here, we can glimpse the Japanese government's reluctant attitude toward social integration. Naturally, the life choices of the people concerned and their selection as a system of the receiving society should be discussed at an inherently different level. In modern society, regardless of the percentage of immigrants who intend to settle permanently, integration policies are required to incorporate immigrants into the political and economic system based on meritocracy.

In this chapter, We compare the recent increasing immigrant population in Japanese society to the Japanese people from the perspective of the theory of stratification, using census data. Unlike previous studies, this chapter is unique in that it simultaneously considers the positions of multiple immigrants in the hierarchy. From the analysis results, we were able to posit that education, nationality, and gender have a multilayered effect on the achievement of immigrants' positions in Japanese society. In considering meritocracy in post-industrial societies, previous studies have focused on the connections between gender and occupation. The results of this study show the need to incorporate the axis of immigration in addition to such relationships. This finding is confirmed by analyzing national data from Japan, where ethnic homogeneity is high. In other words, the need to examine gender, occupation, and immigration with a focus on human capital, as in this study, is increasing in Western countries that have large numbers of immigrants.

The study is not without some limitations. First, the use of only secondary data presents some challenges in the analysis. In capturing the hierarchical structure of Japan, it is desirable to conduct a detailed analysis of the process of status attainment. In addition, in terms of systematic comparison with Western societies, it is necessary to capture the adaptation of the second generation of immigrants in the host society based on the theoretical framework of segmented assimilation. Based on these problematic interests, we need to systematically study Japanese society, focusing on the hierarchical status of immigrants.

Notes

1. However, the refugee status recognition rate is less than 1%, which does not mean that the Japanese government has shifted to an aggressive stance in accepting immigrants..

2. Even before that, there were oldcomers such as Koreans who settled in Japan after World War II, Filipinos who came to Japan as entertainers, and immigrants who stayed through undocumented means. Furthermore, if the Ryukyuan and the Ainu are mentioned, Japanese society was never a mono-ethnic nation.
3. For a summary of migration studies in Japanese urban sociology, see Tani (2013).
4. A few studies of Japanese migrants adopt a labor market and social stratification approach. Examples include Kajita et al. (2005), Takenoshita (2013), Korekawa (2019), Nagayoshi ed (2021), and Ohtsuki (2022).
5. The job change is accepted if the application is approved after application to the government.

Acknowledgements The custom-made tabulation results of the Census of Population (Ministry of Internal Affairs and Communications) in this chapter were provided by the National Center for Statistics, Japan. In addition, we wish to acknowledge the statistical office that provided the underlying data in this research. This research received grants from JSPS: 15K03818 and 18K01966.

References

Alesina, Alberto, Arnaud Devleeschauwer, William Easterly, Sergio Kurlat, and Romain Wacziarg. 2003. Fractionalization. *Journal of Economic Growth* 8 (2): 155–194.

Asahi Newspaper. 2018. Japan's immigration law revised, prime minister acknowledges it is not immigration policy (Nyuukanhoukaisei-Yatou ga Tuikyuu. Shusho-Iminnseisaku denai Ninshiki). *Asahi Newspaper Digital Edition*, 2018, November 5.

Banting, Keith, and Will Kymlicka. 2013. Is there really a retreat from multiculturalism policies? New evidence from the multiculturalism policy index. *Comparative European Politics* 11 (5): 577–598.

Bogardus, Emory S. 1958. Racial distance changes in the United States during the past thirty years. *Sociology and Social Research* 43: 27–135.

Charles, Maria, and David B. Grusky. 2004. *Occupational ghettos*. Stanford: Stanford University Press.

Charles, Maria, and Karen Bradley. 2009. Indulging our gendered selves? Sex segregation by field of study in 44 countries. *American Journal of Sociology* 114 (4): 924–976.

Fisher, Max. 2013. A revealing map of the world's most and least ethnically diverse countries. *The Washington Post*. https://www.washingtonpost.com/news/worldviews/wp/2013/05/16/a-revealing-map-of-the-worlds-most-and-least-ethnically-diverse-countries/. Accessed 3 May 2013.

Goodman, Sara Wallace. 2010. Integration requirements for integration's sake? Identifying, categorising and comparing civic integration policies. *Journal of Ethnic and Migration Studies* 36 (5): 753–772.

Gordon, Milton M. 1964. *Assimilation in American life: The role of race, religion and national origins*. Oxford: Oxford University Press.

Heath, Anthony F., Catherine Rothon, and Elina Kilpi. 2008. The second generation in Western Europe: Education, unemployment, and occupational attainment. *Annual Review of Sociology* 34: 211–235.

Higuchi, Naoto. 2016. "Roudou (Labor)". In *Iminnseisaku toha Nani ka-Nihon no Genjitsu kara Kangaeru* (*What is the immigration policy?—Considering from Japan*), ed. Sachi Takaya, 129–144. Kyoto: Jimbunshoin (in Japanese).

Hollifield, James F., and Michael Sharpe. 2017. Japan as an 'Emerging Migration State.' *International Relations of the Asia-Pacific* 17: 371–400.

Hosokawa, Chihiro. 2018. Kokunaijikeiretsu, Kokusaihikaku niokeru shokugyoubunrui-1980-2010nen Kokuseichousa Kyotsuubunrui no sakusei to ISCO-88 tono taioiu (Classification of occupations for domestic time series and international comparisons creation of a 1980–2010 National Census common classification of occupations and correspondence with ISCO-88), *JSPS KAKEN Grant Research Paper, The Future of the Education Society* (Reports Grant number 15K03818), 58–88 (in Japanese).

Immigration Services Agency. 2023. List of residence qualifications. https://www.moj.go.jp/isa/applications/guide/qaq5.html?. Accessed 6 Jan 2023.

Liu-Farrer, Gracia. 2011. *Labour migration from China to Japan—International students, transnational migrants*. London: Routledge.

Milly, Deborah J. 2014. *New policies for new residents—Immigrants, advocacy, and governance in Japan and beyond*. Ithaca: Cornell University Press.

Kajita, Takamichi. 1988. Ethnicity to Shakaihedo (Ethnicity and social change). Tokyo: Yushindo.

Kajita, Takamichi, Tanno Kiyoto, and Naoto Higuchi. 2005. *Kao no Mienai Teijuuka: Nikkei Brajirujin to Kokka, Shijo, Imin Netowa-ku (Invisible residents—The Japanese Brazilians and state, market and migrant network)*. Nagoya: University of Nagoya Press (in Japanese).

Korekawa, Yu. 2019. *Iminnukeire to Shakaitekitougou no Reality-Gendai Nihon niokeru Iminn no Kaisoutekichii to Shakaigakutekikadai* (*The reality of immigration and social integration-social stratification and sociological issues of immigrants in contemporary Japan*). Tokyo: Keisoshobo (in Japanese).

Nagayoshi, Kikuko. 2020. *Iminn to Nihonshakai-data de Yomitoku Jittaito Syoraizou* (*Immigrants and Japanese society: The actual situation and future vision from data analysis*). Tokyo: Chuokoron-Shinsha (in Japanese).

Nagayoshi, Kikuko (ed.). 2021. *Nihon no Iminntogo—Zenkokuchousa karamiru Genkyo to Shoheki* (*Immigrant integration in Japan—Current status and challenges*). Tokyo: Akashishoten (in Japanese).

Oda, Akeko, Reiko Yamato, and Hiroshi Taroumaru. 2014. Seibetsu-shokuiki-bunri ha chiiki niyotte dou kotonaruka (How do occupational gender segregations differ by region?). In *Higashi-Asia no roudou-shijo to shakai-kaisou* (*Labor market and social stratification in East Asia*), ed. Hiroshi Taroumaru. Kyoto: Kyoto Daigaku Gakujutu Shuppankai (in Japanese).

Ohtsuki, Shigemi. 2022. Shakaikaisou to Iminn (Social stratification and immigrants). In *Kakusa to Bundan/Haijo no shosou wo yomu*(*Considering various aspects of disparity and inequality*), ed. Takuya Hayashi, Shunsuke Tanabe, and Mitsunori Ishida. Kyoto: Koyoshobo (in Japanese).

Oishi, Nana. 2018. Koudojinzai Senmonjinzai wo Meguru Ukeireseisaku no Kansei-Seidotekidoukaka to Genjitsu (The pitfalls of skilled migration policies in Japan). *Japanese Sociological Review* 68 (4): 549–566 (in Japanese).

Okuda, Michihiro, and Junko Tajima. 1991. *Ikebukuro no Ajiakei Gaikokujin-Shakaigaku-teki Jittai houkoku* (*Asian foreigners in Ikebukuro: A report from sociological viewpoint*). Tokyo: Mekong (in Japanese).

Portes, Alejandro, and Min Zhou. 1993. The new second generation: Segmented assimilation and its variants. *The Annals of the American Academy of Political and Social Science* 530: 74–96.

Portes, Alejandro, and Rubén G. Rumbaut. 2001. *Legacies—The story of the immigrant second generation*. California: University of California Press.

Sassen, Saskia. 1991. *The global city: New York, London, Tokyo*. Princeton: Princeton University Press.

Shipper, Apichai W. 2008. *Fighting for foreigners-immigration and its impact on Japanese democracy*. Ithaca: Cornell University Press.

Takahata, Sachi. 2016. Can certified care workers become long-term settlers? Case study of 49 Filipinos under the Japan-Philippines economic partnership agreement. *International Journal of Japanese Sociology* 25: 27–39.

Takenoshita, Hirohisa. 2013. Labour market flexibilisation and the disadvantages of immigrant employment: Japanese-Brazilian immigrants in Japan. *Journal of Ethnic and Migration Studies* 39 (7): 1177–1195.

Tani, Tomio. 2013. Toshi to Ethnicity—Jinkougenshoushakai no Iriguchi ni Tatte (City and ethnicity—at the doorstep of a society in population decline). *The Annals of Japan Association for Urban Sociology* 31: 35–60 (in Japanese).

Van den Berghe, P.L. 1976. Ethnic pluralism in industrial societies—A special case? *Ethnicity* 3: 242–255.

Yamada, Jun. 2018. Why aren't "Foreign Workers" called "Immigrants"? Government and media's "Impression Manipulation" (Naze "Gaikokujinroudousha" wo "Iminn" to yobanainoka? Seifu to Media no "Inshousousa"). *Yahoo Japan news*, 413–423. https://news.yahoo.co.jp/byline/yamadajun/20181028-00102091. Accessed 28 Oct 2018.

Multidimensional Structure of Occupational Segregation in Japan and the United States: Comparison of Trends from 1980 to 2010

Yuya Saitoh

1 Background

In this chapter, we focus on Japan and the U.S. as countries where industrialization has progressed and clarify the trend of occupational segregation from 1980 to 2010. Since 1980, more and more people in Japan and the U.S. have gone on to higher education, raising the level of education in society. In addition, the labor force participation rate of women has also increased, bringing more women into the labor market. Therefore, it is highly likely that some changes have occurred in the factors related to occupational segregation. This chapter focuses on the varying strengths of the effects of gender and education on occupational segregation and will compare occupational segregation in these two countries.

With these transformations in the level of education and labor force participation rates of women in the U.S. and Japan between 1980 and 2010 in mind, let us consider the changing impact of education and gender on occupational segregation. First, the variation in occupational segregation caused by higher education can be considered in two ways. The first is the industrialization theory, which argues that as industrialization progresses, the importance of educational background in achieving professional status increases (Treiman 1970). From the standpoint of industrialization theory, the influence of education on occupational segregation is expected to strengthen because the highly educated group is engaged in occupations with high socioeconomic status that require a high skill level, such as professional and managerial positions, and the occupational segregation from the less educated group becomes more distinct. Second, the importance of education in the labor market decreases as the trend toward higher education increases, and more highly educated workers enter the labor market

Y. Saitoh (✉)
Institute of Social Science, The University of Tokyo, 7-3-1 Hongo, Bunkyo-Ku, Tokyo 113-0033, Japan
e-mail: saito.tmu@iss.u-tokyo.ac.jp

K. Nakao (ed.), *Multidimensional Aspects of Occupational Segregation*, Behaviormetrics: Quantitative Approaches to Human Behavior 18,
https://doi.org/10.1007/978-981-99-8513-5_5

(Jackson et al. 2005). As the relative importance of higher education declines in the labor market, some of the more educated will find themselves in lower skill jobs that, traditionally, would have been occupied by less-educated workers. Consequently, occupational segregation by educational background may be relatively weak.

Occupational segregation by gender can be caused by various factors, but one possible explanation is from the perspective of the welfare regime. There is no consensus on which welfare regime Japan falls into, but Japanese social policy has become more familialistic since the welfare reform around 1980, which has been maintained up to the present (Osawa 2013). Hence, women in Japan tend to take on care work and leave the labor market when they get married or form families. In addition, because Japan has a tax system that favors male-breadwinner households, married women will engage in low-wage, peripheral labor in order to receive tax benefits. Therefore, when married women re-enter the labor market, they are engaged in occupations different from those of men. Thus, the effect of gender on occupational segregation is expected to increase relative to the increase in the labor force participation rate of women in Japan because more women than men will be engaged in different occupations. On the other hand, in the U.S., where the liberal regime has been strengthened since the global economic crisis in 1970, there were no restrictions or support for women's employment (Esping-Andersen 1996, 1999). Consequently, the labor force participation rate of women with children tends to decline, but women as men tend to be employed full-time and are more likely to hold managerial positions than in other welfare regimes (Mandel and Semyonov 2006). Therefore, compared to Japan, the effect of gender on occupational segregation is expected to weaken, as the increase in the labor force participation rate of women is expected to reduce the occupational segregation between men and women.

As industrialization progresses, the criteria for occupational achievement should change from the attribute principle to the performance principle, so the influence of gender on occupational segregation weakens, relatively, while the impact of education strengthens, relatively, between 1980 and 2010. This chapter examines whether these predictions hold true for the U.S. and Japan using census data from these two countries.

2 Previous Research

Occupational segregation by gender is considered to be a factor that explains social inequality, such as the wage gap between men and women, and studies have been conducted in the U.S. for a long time (Petersen and Morgan 1995; Reskin 1993). In the U.S., occupational segregation by gender shrank substantially between 1970 and 1980, as in other developed countries, and has declined slightly or stagnated since then (Blau et al. 2013; Weeden 1998). Blau et al. (2013) examined the change in occupational segregation by gender by decomposing it into occupation mix effects and gender composition effects using the method proposed by Fuchs (1975). The occupational mix effect refers to the change in occupational segregation caused by amendments in the relative size of occupations when the ratio of males (females) in an occupation remains unchanged. Moreover, the gender composition effect indicates the change in occupational segregation caused by the variation in the ratio of males (females) in an occupation when the relative size of the occupation is unchanged. They point out that the reduction in occupational segregation by gender from 1970 to 2009 is caused by both the occupational mix effect and the gender composition effect. In particular, with the gender composition effect, the authors found that the increase in the ratio of women in professional, managerial, sales, and clerical occupations had a significant impact on the ratio of women in "male-dominated occupations," in which the ratio of women was low in 1970. Moreover, with the gender composition effect, the effect of the increase in the ratio of women in professional, managerial, sales, and clerical occupations is clearly significant among stereotypical "male occupations" in which the ratio of women in 1970 was small.

The trend of occupational segregation by gender in Japan has been expanding in recent years (Charles and Grusky 2004; Yamaguchi 2017). Uchikoshi and Mugiyama (2020) contended that these studies did not examine the change in occupational segregation by gender by decomposing it into the occupational mix effect and gender composition effect and analyzed the trend of occupational segregation by gender from 1980 to 2005. The results revealed that occupational segregation by gender in Japan was maintained or decreased slightly from 1980 to 2005. Furthermore, by decomposing the factors of this transformation, the occupational mix effect changes to increase occupational segregation by gender, while the gender composition effect changes to decrease occupational segregation by gender. Hence, the two effects canceled each other out and resulted in less change in occupational segregation by gender in Japan.

In addition, the influence of educational background shows distinctive differences between Japan and the U.S. In the U.S., the trend toward weaker occupational segregation by gender is more pronounced when employees have higher levels of education (Blau et al. 2013; Jacobs 1999). For example, Blau et al (2013) indicate that occupational segregation by gender from 1970 to 2009 changed only by 1.38 percentage points from 60.57% to 59.19% for secondary school graduates but decreased by 21.38 percentage points from 61.73% to 40.35% for college graduates. There has been a large influx of well-educated women into previously male-dominated professional and managerial positions, and segregation has shrunk among the highly educated. On the other hand, the degree of reduced segregation was smaller for the less educated group because women did not flow into construction and production occupations where the proportion of men was higher.

In Japan, Yamaguchi (2017) analyzes the relationship between changes in occupational segregation by gender and educational background and presents results that he calls "the paradox of occupational segregation." This paradox that he refers to is the phenomenon that when human capital, such as age, education, and years of experience, is equalized for men and women, occupational segregation increases, and men and women become unequal. Yamaguchi compared occupational segregation by gender in an actual sample with that in a counterfactual situation in which the human capital of males and females is equalized, using data from Japan in 2005.

His findings reveal that occupational segregation by gender is greater when human capital is equalized than in actual data. There is a tendency for gender segregation to increase in interpersonal professions such as education, medical care, and nursing care, suggesting that higher education does not necessarily lead to a reduction in occupational segregation by gender in Japan.

Hence, two points emerge regarding the trend of occupational segregation in Japan and in the U.S. First, the trend of occupational segregation by gender in Japan and the U.S. is shrinking overall. The degree to which gender segregation is shrinking varies, but the long-term trend is that occupational segregation by gender is diminishing. Second, the relationship between occupational segregation by gender and educational background differs from Japan in the U.S. In the U.S., the influx of highly educated women into professional and managerial positions, where the proportion of men is high, has led to a reduction in gender-based occupational segregation among the highly educated. In Japan, women with higher education are more likely to work in interpersonal professions, where the ratio of women to men is higher, so it is possible that higher education will not lead to a reduction in occupational segregation by gender. Therefore, in the U.S., the influence of gender on occupational segregation is expected to decline after 1980, and the influence of education is expected to become stronger. Although occupational segregation by gender is reducing in Japan, it is quite possible that the effect of gender on occupational segregation is greater than the effect of education, given that women are becoming more educated.

In previous studies, occupational segregation by gender was analyzed using dissimilarity indices. Since changes in dissimilarity indices can be decomposed into changes in occupational mix and changes in the gender composition of occupations, it is possible to study the trend of occupational segregation by gender in detail. However, occupational segregation in the labor market can be caused not only by gender but by education, age, and other factors. In addition, it is not possible to clarify which of the factors that create occupational segregation in the labor market are the major factors, or whether the major factors have changed over time by simply examining occupational segregation by gender. Based on the above, this paper focuses on gender and education and analyzes the occupational segregation created by these factors as a multidimensional structure through correspondence analysis. By analyzing the relationship between attributes and occupation as a multidimensional structure and analyzing the trends, it is possible to discuss the relationship between occupational segregation by gender and occupational segregation by educational background, and how they have changed.

3 Methods and Data

3.1 Analytical Framework

In this chapter, we apply correspondence analysis to a contingency table consisting of occupations and attributes. Correspondence analysis represents the distance between each category of row and column variables in space. As evident in the next section, this chapter analyzes a two-dimensional space as the coordinate space and plots each category of occupation and attribute in that two-dimensional space. Based on the positional relationships of each category of row and column variables plotted in a two-dimensional space, we grasp the characteristics and trends of the multidimensional structure of occupational segregation in Japan and the U.S. For comparison using correspondence analysis, there are two methods: one is to analyze the space common to all time points and verify the changes between time points in that space, and the other is to perform correspondence analysis on the data at each time point, analyze the space for each time point, and compare the results. In the former case, it is assumed that the dimensional structure of the space to be deposited does not change significantly. Since the purpose of this chapter is to understand the characteristics of the dimensional structure analyzed from the contingency table of occupations and attributes for Japan and the U.S. and to understand the trend of the changes in the dimensional structure, we adopt the latter method and confirm whether there are significant changes in the dimensional structure analyzed.

We compare Japan and the U.S. with common categories for attributes but different categories for occupations in them. One method is to convert the occupational classifications of Japan and the U.S. into occupational classifications suitable for international comparison, such as in the International Standard Classification of Occupations (ISCO), and analyze them, but it has been pointed out that the Japanese occupational classification system is very different from that of ISCO (Tanabe 2006). For example, ISCO classifies people as "Professionals" and "Technicians and Associate Professionals" according to their skill level, but in the Japanese occupational classification, there is no distinction between "Professionals" and "Technicians and Associate Professionals," and they are combined into one category as "Professional and Technical Occupations." Tanabe argues that such differences in occupational classifications affect the results of the analysis; therefore, this chapter does not use occupational classifications suitable for international comparisons, such as ISCO.

Another method is to create a comparative occupational classification by mapping the Japanese occupational classification to the U.S.'s, but as with the method of converting to ISCO, there are significant differences between the Japanese and U.S. occupational classifications. Comparing the number of minor occupational categories, the 2010 Japanese occupational classification was 231, while the 2010 U.S. occupational classification was 487. If a common occupational classification were to be created, the occupations in the U.S., which have numerous minor categories, would be matched to the Japanese occupational classification, which has fewer minor categories, but this method is also not appropriate because the Japanese occupational classification reflects aspects of Japanese employment practices.

In some cases, it may be more desirable to use ISCO or to create a common occupational classification for Japan and the U.S. to compare the two countries. The dimensional space deposited from attributes and occupations can be understood as the relative position of each attribute in the labor market. However, to capture the characteristics and trends of the labor market in each country, it is more appropriate to use occupational classifications that reflect the uniqueness of the labor market in each country. For this reason, this chapter uses different occupation codes for the U.S. and Japan. However, to grasp the trend from 1980 to 2010, common occupational categories were created for Japan and the U.S., respectively, and used in the analysis.

3.2 Data

The data for Japan are from the National Population Census in Japan conducted by the Ministry of Internal Affairs and Communications. Since the census is conducted every 5 years, but education is assessed every 10 years, the 1980, 1990, 2000, and 2010 censuses will be used in the analysis. The data for the four census surveys used in this study were obtained using the "order-made summarization" service provided by the National Statistics Center. The "order-made summarization" is a fee-based service that provides data from existing government statistics to meet the needs of users, limited to academic research of high public interest. The data provided are not a subsample extracted from the census, but a contingency table created using the entire census sample. The data for the four time points used in this study were provided as a contingency table consisting of occupation and attributes combining gender, age, and education.

The data for the U.S. are from the Census of Population and Housing and the American Community Survey. This survey is conducted every 10 years and uses a 5% subsample of data from 1980, 1990, and 2000. The 2010 Census of Population and Housing data is also available as a Public Use Micro Sample (PUMS) on the Census Bureau's website, but it does not include variables such as occupation and education that are necessary for analysis. For this reason, we used a 1% subsample of the American Community Survey data for 2010. All data for these four time points were obtained through the Integrated Public Use Microdata Series, International (IPUMS-International).

Moreover, for occupation, as described in the analytical framework, we use common occupational classifications for each of Japan and the U.S. across the four time points. The occupational categories used in the Japanese census were 292 categories in 1980, 293 in 1990, 292 in 2000, and 231 in 2010. The occupational categories at the four time points were minimally integrated to create an occupational classification with 157 categories. The occupational categories used in the U.S. Census of Population and Housing and the American Community Survey were 503 categories in 1980, 502 in 1990, 471 in 2000, and 487 in 2010 (excluding military jobs). For U.S. occupations, minimal category integration was performed between points in time, resulting in 331 categories of occupational classification.

Attributes were created by combining gender, age, and educational background. The genders were divided into two categories: male and female; the ages were divided into four categories: 20s, 30s, 40s, and 50s; the educational background was divided into four categories: junior high school or less, high school, junior college or technical college, and university or more. These three variables were combined to create a variable called attribute with a total of 32 categories, which were used in the analysis.[1]

4 Analysis Results

4.1 Determining the Number of Dimensions

Correspondence analysis was conducted on eight contingency tables from 1980 to 2010 for Japan and the U.S. Before interpreting the appendix, we checked the eigenvalues and cumulative contributions for the dimensions. The eigenvalues for the U.S. shown in the upper left of Fig. 1 drop sharply to the second dimension for all years and gradually decrease after the third dimension, so the two-dimensional solution is adopted for the U.S. as the result of correspondence analysis. The cumulative contribution rates up to the second dimension for the U.S., indicated in the upper right, are 63.03%, 64.73%, 64.38%, and 64.97%.

The eigenvalues and cumulative contributions for Japan are illustrated in the lower part of Fig. 1. The eigenvalues for Japan also dropped sharply to the second dimension for all years and gradually decreased after the third dimension. Therefore, Japan, like the U.S., adopts a two-dimensional solution as a result of correspondence analysis. The cumulative contribution rates for up to two dimensions for Japan are 52.13%, 56.17%, 60.66%, and 62.61%.

4.2 The Structure and Trends of Occupational Segregation in the U.S.

Figure 2 shows the configuration of attributes from 1980 to 2010 in the U.S. First, in the configuration of attributes for 1980 in the upper left of Fig. 2, males are represented in the upper left of the figure and females are in the lower right. Since gender segregation exists at all levels of education, one can infer that people with the same education but different genders have different occupations. The order of education is from top right to bottom left, from university to junior high school. A space between university and junior college for both men and women confirms that there is a difference in the occupations they engage in between university and junior college. Thus, the meaning of the first and second dimensions cannot be clearly

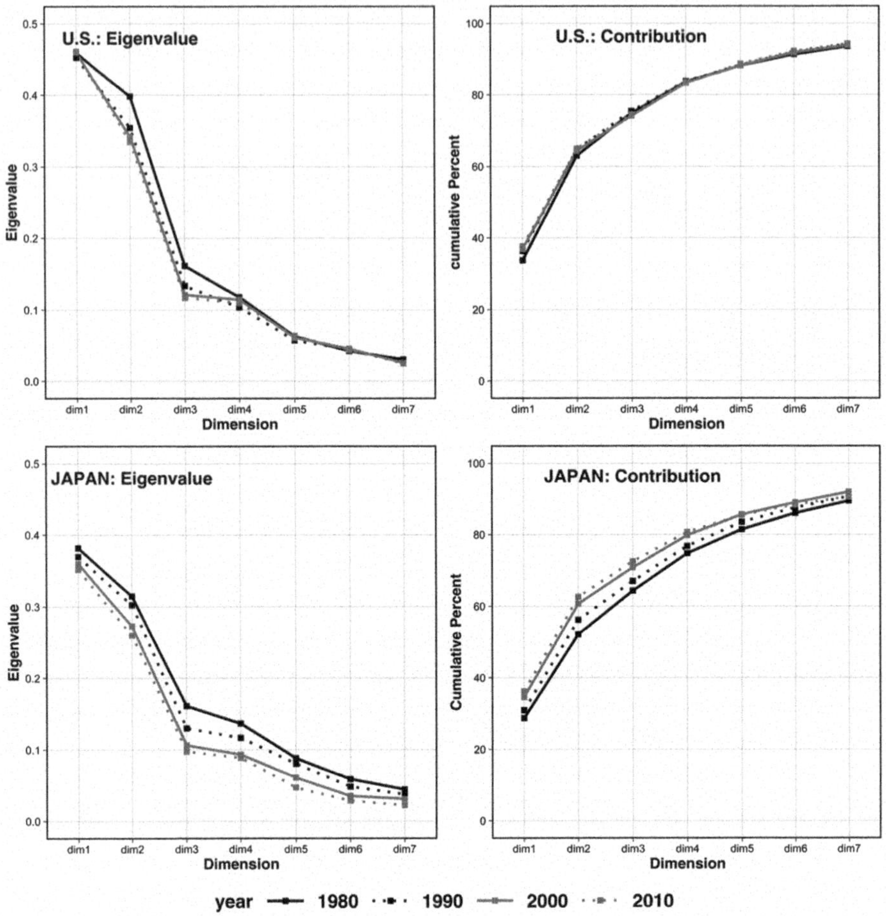

Fig. 1 Eigenvalues and cumulative contribution of Japan and the U.S.

interpreted as of 1980, but occupational segregation is thought to be mainly composed of gender and education.

When we look at the post-1990 configuration of attributes, we see that while the relative positional relationships in the 1980 configuration are generally maintained, there is an overall shift to a rightward rotation, allowing for a clearer interpretation of the primary and secondary axes of the configuration. The first dimension can be interpreted as segregation by education, because in the 2010 configuration of the attributes, the order of education is from right to left of the apposition, with some exceptions. Except for female university graduates, males are located at the top of the figure and females at the bottom, so the second axis can be interpreted as segregation by gender.

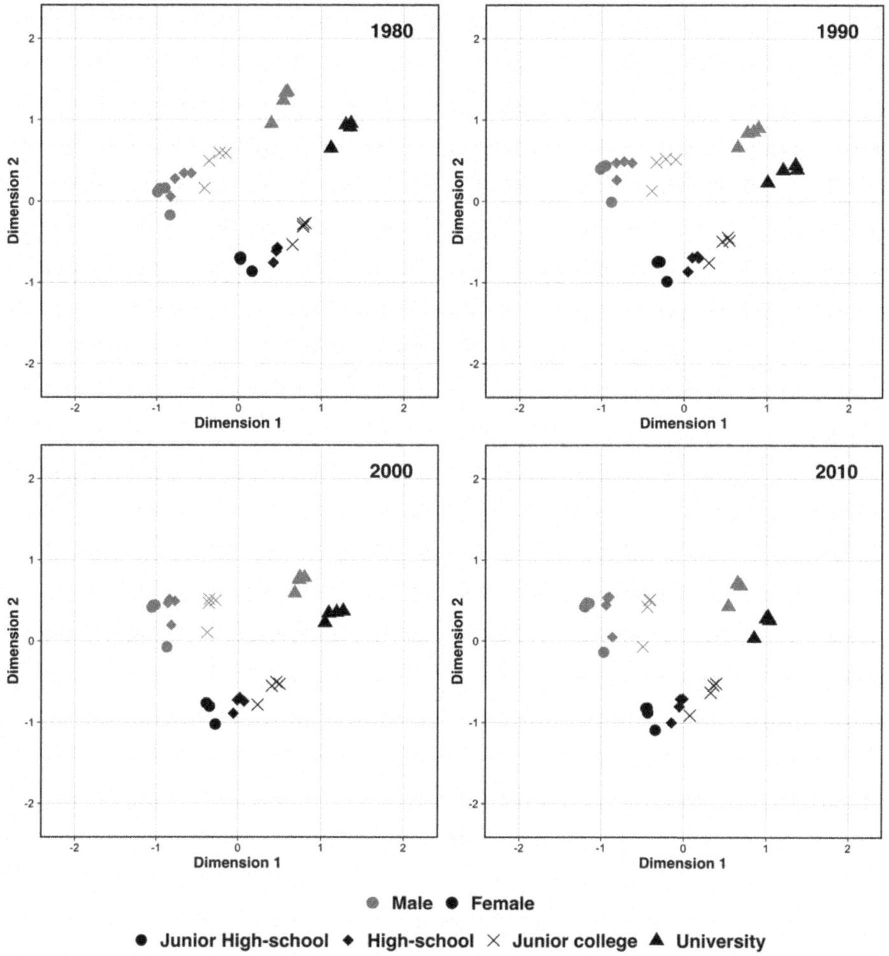

Fig. 2 Configuration of attributes in the U.S. from 1980 to 2010

In addition, although it is a minor change, a comparison of the position of university graduates in the configuration shows that the men and women in the university-educated group were relatively closer in 2010 than in 1980. Since correspondence analysis was applied to different contingency tables, it is not possible to directly compare long and short distances, but the results suggest that in the U.S., in recent years, university graduates may have started to work in relatively similar occupations, regardless of gender. This is consistent with the tendency pointed out in previous studies that occupational segregation is shrinking among university graduates.

Moreover, we interpret the configuration of occupations in the U.S., based on the configuration of attributes as demonstrated in Fig. 3. Using the 1980 configuration as an example, we confirmed the position of the occupation. The letters in the figure refer to major occupational categories. In the 1980 appendix, professionals (A) and managers (B) are mostly located in the first quadrant, and university graduates of both genders are employed in professional and managerial positions. Most of the clerical jobs (C) are in the lower part of the figure, and most of them are in the fourth quadrant. Among women, those who graduated from junior college, high school, and junior high school tend to work in clerical jobs. It is difficult to identify a clear trend in the sales occupation (D), but it is divided into those located at the top of the figure and those located at the bottom of the figure. The sales occupations at the top of the figure are those that sell business-related services and personal services, such as "Insurance sales occupations" and "Securities and financial services sales occupations." The sales occupations in the upper part of the figure are "Insurance sales occupations" and "Securities and financial services sales occupations," and correspond to male university graduates and junior college graduates.

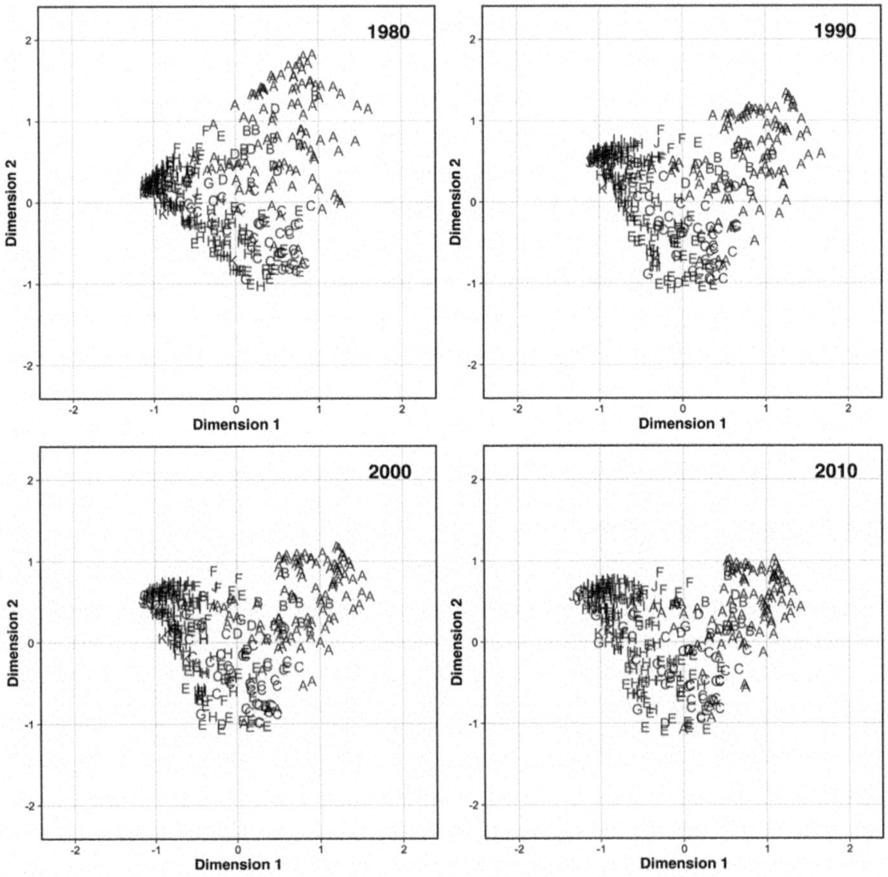

Fig. 3 Configuration of occupation in the U.S. from 1980 to 2010

On the other hand, the sales occupations in the lower right of the figure include "Cashiers" and "Retail salespersons," who are responsible for sales work in retail stores and correspond to female junior colleges, high schools, and junior high schools. In the case of salespersons, occupation content differs by gender. The service occupations (E) are located at the bottom of the figure and correspond to female high school and junior high school. In contrast, most security positions (F) occupy the second quadrant of the configuration and are closely proximate to male junior colleges and high schools. Agricultural workers (G) are in a dense plot of occupations on the left side of the configuration. Other occupations in this dense area include transportation (I), construction (J), and some manufacturing (H), and these occupations are occupied by male high school and middle school graduates. Manufacturing (H) is in the lower part of the figure and in the second quadrant, where the occupational plots are densely arranged and tend to be dominated by male and female high school and middle school graduates. Among manufacturing occupations, those that align with male high school and junior high school students are metalworking and heavy industrial manufacturing, such as automobile mechanics and metal plating machine operators. There are many occupations such as textile sewing machine operators and shoe machine operators, engaged in the manufacture of light industrial products, closely proximate to female high schools and junior high schools. Moreover, occupations involving carrying, cleaning, packaging, and related work (K) are positioned close to male and female high schools and junior high schools.

Because the configuration of occupation corresponds to the configuration of attributes, the axis that constitutes the space cannot be clearly interpreted considering the 1980 configuration of occupation. However, male-dominant occupations, such as transportation, construction, manufacturing of metal products, and heavy industry products, and professional occupations such as "Physicists and astronomers" and "Physicians," are located in the upper left-hand corner of the configuration corresponding to males. In the lower right-hand corner, female-dominant occupations include office work, service occupations, manufacturing of light industrial products such as clothing, and professional occupations such as "Teachers, elementary school" and "Occupational therapists." These are the positions of female-dominant occupations. Professional managers occupy the upper right of the figure, and other occupations are found in its lower left. From the occupational classification, one can observe that occupational segregation consists of the segregation of "male-dominant occupation" and "female-dominant occupation" based on the gender composition ratio in the occupation, and segregation by occupational specialty, complexity, or skill level. It is thought that the disparity between the two groups is due to their specialization, complexity, or skill level.

Furthermore, similar to the configuration of attributes, the occupation has rotated in its entirety from 1980 to 2010, maintaining its relative position. Figure 3 shows that since 1990, the first dimension of configuration of occupation has gradually shifted to the segregation of professional and managerial occupations from other occupations, while the second dimension has changed to the segregation between "female-dominant occupation" and "male-dominant occupation." By 2010, education and occupational specialization/skill level were considered the key factors in occupational segregation in the U.S., while gender and occupational segregation based on gender composition were considered relatively less important. In addition, the number of professional and managerial positions became denser when comparing 1980 and 2010. This is considered a change corresponding to the proximity of men and women in the university-educated group in the configuration of attributes.

In addition, the contribution rate of the first dimension has consistently increased from 34.47% to 38.01% from 1980 to 2010, while the contribution rate of the second dimension continued to decrease, albeit slightly, from 29.26% to 27.31%. Given that the first axis has changed to segregation by education and the second axis to gender over the past 30 years, the primary factor of occupational segregation in the U.S. has changed to education, and its influence has become relatively strong in recent years. Conversely, the second factor for occupational segregation shifted to gender by 2010, and its impact was relatively weaker than that of education.

4.3 The Structure and Trends of Occupational Segregation in Japan

In the results of the correspondence analysis for Japan, we start with the configuration of the attributes. Figure 4 shows the configuration of attributes in Japan from 1980 to 2010. In the 1980 configuration, university graduates of both genders and junior colleges occupy the upper right of the figure, and high school and junior high school are located in a row to the lower left. Males occupy the upper left of the configuration, while females are positioned at the lower right. Compared to the configuration of attributes in the U.S. and in Japan in 1980, the configuration of attributes is characterized by the presence of a series of educational levels. In the U.S., there is a space between university graduates and non-university graduates, and there is a high likelihood of different occupations between these educational backgrounds, whereas, in Japan, university graduates and junior college graduates are likely to have similar occupations, while junior college and high school graduates are likely to have similar occupations. Although there are differences regarding the position of education, the configuration of attributes in Japan in 1980 is relatively similar to in the U.S. in 1980. Therefore, it is difficult to interpret the dimensions clearly, but it is possible to conclude that segregation occurs predominantly by gender and education.

The changes since 1990 indicate both overall and localized changes in apposition have occurred. First, the overall configuration changed from 1980 to 2010, with males on the left side of Fig. 4 and females on the right. The first dimension of 2010 can be clearly interpreted as gender segregation, whereas the second dimension can be said to be segregated by education, as university graduates are located at the top of the figure as of 2010, while high school and junior high school graduates occupy the bottom thereof. In contrast to the results of the U.S. analysis, in Japan, segregation by education became the second dimension, while segregation by gender became the

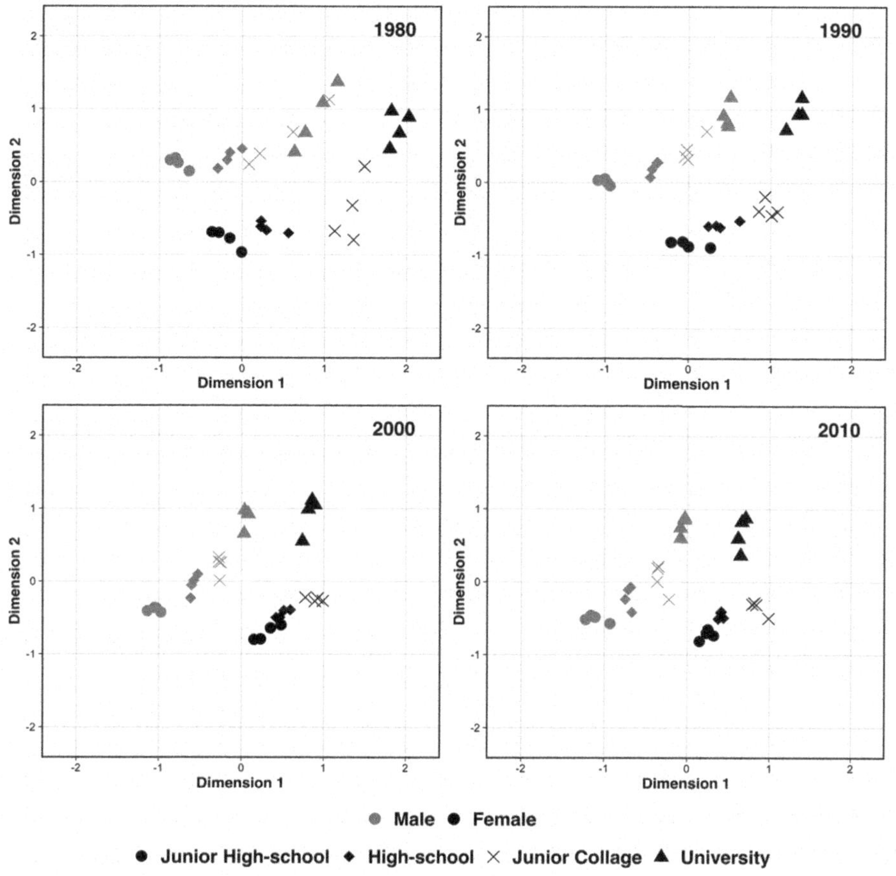

Fig. 4 Configuration of attributes in Japan from 1980 to 2010

first dimension between 1980 and 2010. This means that the influence of gender on occupational segregation has increased the past 30 years.

In addition to the overall change, a local change has occurred in the post-1990 configuration, where women's education is no longer continuously located. Since 1990, university and junior college have moved apart, with junior college moving rather closer to high school. In addition, by 2000 and 2010, female junior colleges were somewhat farther away from high schools and further to the right of the configuration. In Japan, women with a junior college degree were associated with different occupations than women with any other educational background in 2010. Thus, although both overall and local changes in the configuration of attributes have occurred, it can be said that the components of segregation are consistent in that they are composed of education and gender.

We then interpret the configuration of occupations in Japan. Figure 5 shows the configuration of occupations from 1980 to 2010. As in the U.S., the letters in the figure correspond to the major occupational categories. In 1980, professionals (A) were scattered up and down on the left side of the space, indicating that there was a large gender segregation within the professions. Among the professionals, "Doctors" and "Judges, public prosecutors, lawyers" are located at the top of the figure, while "Nurses" and "Childcare workers" are located at the bottom. Managers (B) occupy the upper part of the figure, corresponding to men who graduated from university or junior college. Clerical workers (C) are located from near the origin to the fourth quadrant, and "Transport clerical workers" are close to male high school graduates. "Accountancy clerks" and "Office appliance operators" are close to female high school graduates, so they tend to engage in different office jobs depending on their gender. Most of the sales occupations (D) are located near the origin and slightly to the left of the figure. One can observe that service occupations (E) are more prevalent in the fourth quadrant from the origin and are associated with women who have graduated from high school or junior high school. Security occupations (F), transportation (I), and construction (J) are mostly associated with male high school and junior high school because they form dense plots of occupations in the second quadrant. Agriculture (G) and manufacturing (H) are in the second and third quadrants, close to junior high school graduates. Occupations involving carrying, cleaning, packaging, and related work (K) are positioned in the second to third quadrant, as are agriculture (G) and manufacturing (H), and are associated with junior high school graduates.

In 1980, professional and managerial occupations were separated from other occupations in the upper right and lower left, corresponding to the educational background, and workers with a university degree were engaged in highly specialized occupations. On the other hand, there are many male-dominant occupations in the upper left of the figure and many female-dominant occupations in the lower right, which can be considered as occupational segregation by gender. This configuration of occupation had undergone both overall and local changes between 1980 and 2010, corresponding to the configuration of attributes; in the configuration of occupation in 2010, the segregation of male-dominant occupation and female-dominant occupation occurred left to right, respectively, and the segregation of occupation by specialty and skill level occurred up and down. In addition, corresponding to the local changes in female junior colleges identified in the configuration of attributes, some professionals are located above 1.0 point in the first dimension in the 2010 configuration of occupation. These occupations are "Kindergarten teachers," "Childcare workers," "Midwives," "Nutritionists," "Public health nurses," and "Nurses."

Changes in the contribution rate revealed that the first dimensions was 28.69% in 1980, which increased to 36.14% in 2010. On the other hand, the contribution of the second dimension increased from 23.44% to 26.46% over 30 years. Given that the first dimension of occupational segregation in Japan in 2010 was based on gender, the most crucial factor impacting occupational segregation in Japan has shifted to gender, and its effect has increased in recent years. While the influence of education has strengthened over the past 30 years, it has become less important than gender.

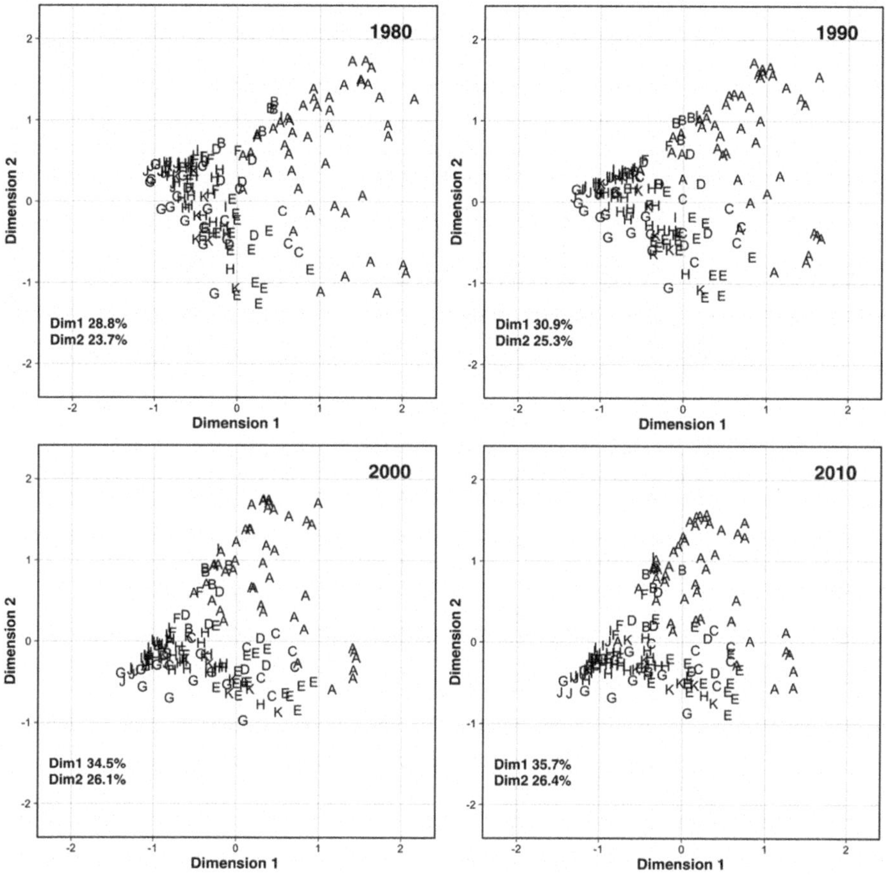

Fig. 5 Configuration of occupation in Japan from 1980 to 2010

5 Discussion

To summarize the results of the analysis thus far, occupational segregation in the U.S. and Japan in 1980 mainly consists of gender and education, and furthermore, it is common that the first and second dimensions cannot be clearly interpreted due to the influence of either gender or education. However, it has changed completely between 1980 and 2010. In the U.S., from 1980 to 2010, the first dimension of the configuration analyzed by correspondence analysis changed to education, and the second dimension changed to gender. The contribution of the first dimension has increased consistently since 1980 so that the influence of education on occupational segregation has gradually become stronger, while the contribution of the second dimension has decreased slightly so that the influence of gender has weakened in recent years. During the past three decades, the U.S. has had an increase in the number of highly educated workers, but the relative importance of education has not been undermined, and the trend has not been as predicted by Jackson et al. (2005). In addition, the occupational segregation between men and women has tended to diminish among university-educated men and women, although a clear segregation between men and women has been maintained among the less educated. The change in occupational segregation in the U.S., with the declining influence of gender and the increasing influence of education, is consistent with the predictions of industrialization theory.

In contrast, in Japan, from 1980 to 2010, the first dimension of occupational segregation shifted to gender and the second dimension to education. Although previous studies have pointed out that occupational segregation by gender in Japan has tended to dissipate in recent years, the contribution rate of the first dimension has consistently increased over the past 30 years, and the effect of gender has relatively increased in recent years when analyzed considering education and age. However, this result does not indicate that the relative importance of education has decreased. This is because the contribution rate of the second dimension has also been consistently increasing from 1980 to 2010. Furthermore, given that the ratio of workers with university degrees in the labor market is higher in the U.S. than in Japan, the results for Japan are not as predicted by Jackson et al. (2005). Thus, although the importance of educational background for occupational segregation seems to be increasing in Japan, as predicted by the industrialization theory, the importance of gender for occupational segregation is increasing even more. Since occupational segregation in Japan is characterized by gender segregation in the labor market, the industrialization theory alone cannot explain the trend of occupational segregation.

It is worth noting that although both the U.S. and Japan have experienced increases in women's labor force participation since 1980, very different changes have occurred in occupational segregation in the U.S. and Japan. As already discussed in the previous chapters, in Japan, most female workers are unmarried full-time workers and married part-time workers, but the increase in female labor force participation rates from 1980 to 2010 is largely due to the increase in part-time workers. This may be because Japan has a tax and social security system that discourages married women from entering the workforce, and because women are subject to certain restrictions in their choice of occupation before they can re-enter the labor market, many of them become part-time workers. However, the Japanese tax and social security systems are designed to suppress married women's income by favoring male-headed households, so married women keep their income low by adjusting their occupation and working hours. These systems are based on the principle of familialism, which holds that women should be responsible for housework, childcare, nursing care, and other care work within the household. Of course, in Japan too, laws have been put in place since 1980 to encourage women to work, such as the Basic Act for Gender Equal Society and the Childcare Leave Law, but their impact on increasing the number of women in full-time work is insufficient. In other words, the occupational segregation of men and women could be shaped by Japan's familialistic social policy.[2]

Thus, occupational segregation by gender in Japan is hardly the result of free individual choice, but may be the result of social policies restricting women's freedom to choose their occupations. As previous studies have pointed out, gender segregation is often viewed as the result of free individual choice or natural selection based on gender-specific characteristics, but in reality, it is formed by sexist institutions and stereotypes about gender differences (Charles and Bradley 2009). Furthermore, as these studies warn, gender segregation reinforces gender stereotypes and prevents women from recognizing that their life chances are limited.

In this chapter, we have compared Japan and the U.S. as industrialized countries. If the changes that have occurred in the U.S. are typical of the changes that occur in industrialized countries, Japan has shown changes that are not typical. What the analysis suggests is that women's re-entry into the labor market itself and their occupational choices when re-entering the labor market are limited, and their tendency to engage in care work even within the labor market, may be the result of Japan's strong familialistc social policies. Since familialism is discussed as a characteristic of East Asian countries including Japan, it is necessary to verify with data whether the characteristics of occupational segregation in Japan confirmed in this chapter are also confirmed in occupational segregation in East Asian countries. It is also necessary to further clarify the characteristics of occupational segregation in Japan by comparing not only East Asian countries but also other Asian and European countries. The next chapter and the following chapters will deal with these issues.

Notes

1. In this paper, we do not include race as an attribute variable for the U.S., as we will use the same variables as in Japan in our analysis. Using data from the U.S., we added race to the list of attributes, and confirmed that race has a smaller impact relatively on occupational segregation as education and gender.
2. Of course, since 1980, the Long-Term Care Insurance Act was enacted in Japan, and the aging of the population has also had an impact, increasing the demand for labor to provide nursing care, and many women have been engaged in such nursing care. Previous studies have indicated that the expansion or contraction of demand for such industries and occupations affects occupational segregation.

Acknowledgements This research received grants from JSPS: 15K03818 and 18K01966. Data for the United States were provided by IPUMS (https://www.ipums.org/). In addition, we wish to acknowledge the statistical offices that provided the underlying data making this research possible: Bureau of the Census, United States. The custom-made tabulation results of the Census of Population (Ministry of Internal Affairs and Communications) in Japan were provided by the National Center for Statistics, Japan.

References

Blau, Francine D., Peter Brummund, and Albert Yung-Hsu. Liu. 2013. Trends in occupational segregation by gender 1970–2009: Adjusting for the impact of changes in the occupational coding system. *Demography* 50: 471–492.

Charles, Maria, and David B. Grusky. 2004. *Occupational ghettos: The worldwide segregation of women and men*. Stanford: Stanford University Press.
Charles, Maria, and Karen Bradley. 2009. Indulging our gendered selves? Sex segregation by field of study in 44 countries. *American Journal of Sociology* 114 (4): 924–976.
Esping-Andersen, Gusta. 1996. *Welfare states in transition: National adaptations in global economies*. London: SAGE Publications.
Esping-Andersen, Gusta. 1999. *Social foundations of postindurstrial economies*. NewYork: Oxford University Press.
Fuchs, Victor R. 1975. A note on sex segregation in professional occupations. *Explorations in Economic Research* 2: 105–111.
Jacobs, J.A. 1999. The sex segregation of occupations: Prospects for the 21st century. In *Handbook of gender and work*, ed. C.A. Newbury Park, 125–141. Sage Publications.
Jackson, Michell, and H. GoldthorpeJohn, and Mills, Colin,. 2005. Education, employers and class mobility. *Research in Social Stratification and Mobility* 23: 3–33.
Mandel, Hadas, and Moshe Semyonov. 2006. A welfare state paradox: State interventions and women's employment opportunities in 22 countries. *American Journal of Sociology* 111: 1910–1949.
Osawa, Mari. 2013. *Seikatsu hosho no Governance: Gender to okane no nagare de yomioku* (*Governance of livelihood security*). Yuhikaku (in Japanese).
Petersen, Trond, and Laurie A. Morgan. 1995. Separate and unequal: Occupation-establishment sex segregation and the gender wage gap. *American Journal of Sociology* 101: 329–365.
Reskin, Barbara. 1993. Sex segregation in the workplace. *Annual Review of Sociology* 19: 241–270.
Tanabe, Shunsuke. 2006. ISCO to SSM shokugyou bunrui no souiten no kentou: Kokusai hikaku chousa ni okeru shokugyou data ni kansuru kenkyu note (Differences in occupational classifications, ISCO and SSM: Note on the use of occupational information in international comparative research). *Sociological Reflections* 27: 53–78 (in Japanese).
Treiman, Donald J. 1970. Industrialization and social stratification. *Sociological Inquiry* 40: 207–234.
Uchikoshi, Fumiya, and Ryota Mugiyama. 2020. Nihon ni okeru seibetsu-shokuiki-bunri no susei (Trends in occupational sex segregation in Japan: A decomposition analysis of census data, 1980–2005). *Journal of Population Studies* 56: 9–23 (in Japanese).
Weeden, Kim A. 1998. Revisiting occupational sex segregation in the United States, 1910–1990: Results from a log-linear approach. *Demography* 35: 475–487.
Yamaguchi, Kazuo. 2017. *Hatarakikata no Fubyoudou: riron to jisshou bunseki* (*Gender Inequalities in the Japanese Workplace and Employment: Theories and Empirical Evidence*). Nikkeibook (in Japanese).

Comparison of Occupational Segregation in East Asia: Analyzing the Structure of Segregation by Educational Background and Gender

Yuki Kuwana

1 Introduction

As industrialization progressed, it was expected that the criteria to attain occupational status would shift from ascription to achievement (Treiman 1970). Therefore, it was predicted that the relationship between educational experience and occupational status would strengthen. However, the signaling effect of education on occupational status has been declining in some occupations because of an increase in education and changes in industrial structure (Jackson et al. 2005). In the 2000s in Japan, the correlation coefficient between years of education and occupational prestige scores tended to decline among young people (Nakao 2018).

Our project commenced with the question of why the relationship between education and occupational prestige scores is declining in Japan, a country that is becoming increasingly industrialized and should be moving from an attribution-based to a performance-based system. We found that the effect of education on occupation is diminishing, especially for women who began their first job after 1991 (Wakita 2018), the beginning of the long-term recessionary period. Not only to the linear relationship between years of education and occupational prestige scores but also clarified the relationship between occupational and educational categories in greater detail by comparing historical and national contexts.

Why does the distance between occupations by gender appear so clearly in Japan? In this chapter, we compare Japan with East Asian countries (Korea and Taiwan) to elucidate this question. In chapter "An International Comparison ofOccupational Segregation by Gender: Through a Comparison of 14 Countries", an international comparison is made for various countries using data from the Integrated Public

Y. Kuwana (✉)
Faculty of Political Science and Economics, Kokushikan University, 4-28-1 Setagaya, Setagaya-Ku, Tokyo 154-8515, Japan
e-mail: k.yuki@kokushikan.ac.jp

K. Nakao (ed.), *Multidimensional Aspects of Occupational Segregation*,
Behaviormetrics: Quantitative Approaches to Human Behavior 18,
https://doi.org/10.1007/978-981-99-8513-5_6

Use Microdata Series, International (IPUMS-International). In chapter "Multidimensional Structure of Occupational Segregation in Japan and the United States: Comparison of Trends from 1980 to 2010", comparisons between Japan and the U.S. were examined over a longer time interval. However, these chapters do not include East Asian countries in their analysis.

East Asian countries have similar structural conditions that shape of social stratification, such as industrialization centered on labor-intensive manufacturing, thorough agrarian reform, and universal and rapid educational expansion (Arita 2009). Furthermore, there are similarities in factors that seem to maintain gender segregation, such as changes in industrial structure and a family-oriented social welfare system (Brinton 2001; Chang 2010). However, there are differences in institutional conditions such as labor market practices and education systems (Arita 2009). For example, Taiwan is unique among East Asian countries in that it has a high female employment rate.

Thus, in clarifying occupational segregation in Japan, East Asian countries are important comparators with interesting similarities and differences. Nevertheless, the fact that East Asian countries have not been the subject of analysis in other chapters is mainly due to the availability of datasets. The statistical bureaus of Korea, Taiwan, and Japan do not provide census data to IPUMS, nor do they publish analyzable contingency tables. This necessitated the need to make East Asian comparisons using comparable data sets separately from the other chapters. This chapter identifies similarities and differences in occupational segregation by gender and education between other East Asian countries and Japan.

2 Previous Research

2.1 Changes in Industrial Structure and Gender Segregation in East Asian Countries

A point that is often mentioned in explaining gender occupational segregation in Japan is the fact that gender occupational segregation clearly remains when compared to Western Europe. Gender-based occupational segregation tends to be reduced in the U.S. (Jacobs and Lim 1992), but it still tends to be maintained in Japan. However, gender occupational segregation from all perspectives does not remain in Japan. It was pointed out that while overall occupational segregation is expanding in Japan, controlling for changes in occupational distribution, segregation by gender composition within occupations is shrinking (Charles and Grusky 2004). It was also pointed out that in Japan, unlike Western European countries, many women work as part-timers in the manufacturing industry, thereby reducing occupational segregation compared to the UK and the U.S. (Shirahase and Ishida 1994). On the other hand, a study that examined the long-term trend of occupational segregation by gender in terms of changes in the occupational structure and the ratio of men to women in occupations found that women continued to be disproportionately represented in occupations such as office workers, service workers, nurses, child care workers, and some health care workers, where the proportion of women was originally large, and that as a result there was a tendency for gender segregation in the occupational structure in Japan to increase. As a result, a trend toward greater gender segregation in the occupational structure in Japan was confirmed (Uchikoshi and Mugiyama 2020). As shown above, the gender segregation of occupations in Japan has been different from that in Western Europe and the United States. In particular, the lack of increase in the proportion of male workers in some professional and service occupations is a unique feature. These studies suggest that either there has been no change in the industrial structure in Japan as in Western Europe, or that even if the composition of major industries has changed, there has been little change in the gender composition within occupations.

Therefore, let us examine the relationship between changes in industrial structure and changes in occupational segregation in Western Europe and consider how this differs from East Asia, including Japan: In an international comparative study by Goldin (1994), in societies where agriculture is a major industry, women tend to work as family employees, and women's labor force participation rates tend to be high. In societies where manufacturing is the main industry instead of agriculture, men tend to be employed as manual laborers, while women are often unemployed as full-time housewives. Therefore, the labor force participation rate of women in such societies is low. On the other hand, in societies where the service sector has replaced manufacturing as the main industry, women tend to enter the labor market as non-manual workers compared to men. Therefore, the labor force participation rate of women tends to be higher than in societies where the manufacturing sector is the main industry. Goldin calls this change in the labor force participation rate of women depending on the stage of industrial structure a "U-shape" transition. In advanced Western European countries, the shift in industrial structure from agriculture to manufacturing and from manufacturing to the service sector occurred earlier and over a longer period than in other regions.

However, in East Asian countries, including Japan, the change from an agriculture-based to a manufacturing-based industry occurred only after World War II, and the time until the expansion of the service sector was short thereafter. Compared to advanced Western European countries, East Asian countries still have a high percentage of their working population engaged in agriculture (Fig. 1), and a low percentage of their working population engaged in the service sector (Fig. 2). The common trends in industrial structure in East Asia are characterized by a larger agricultural population and a smaller service sector than in Western Europe. This is a condition common to all East Asian countries.

Therefore, if East Asia is commonly experiencing a gender segregation transition that differs from that of Western Europe, it may be due to a change in industrial structure that differs from that of Western Europe. However, it is still unclear to what extent and in what ways these conditions are common across East Asia. Therefore, in the next and subsequent sections, we will examine East Asian differences in women's employment.

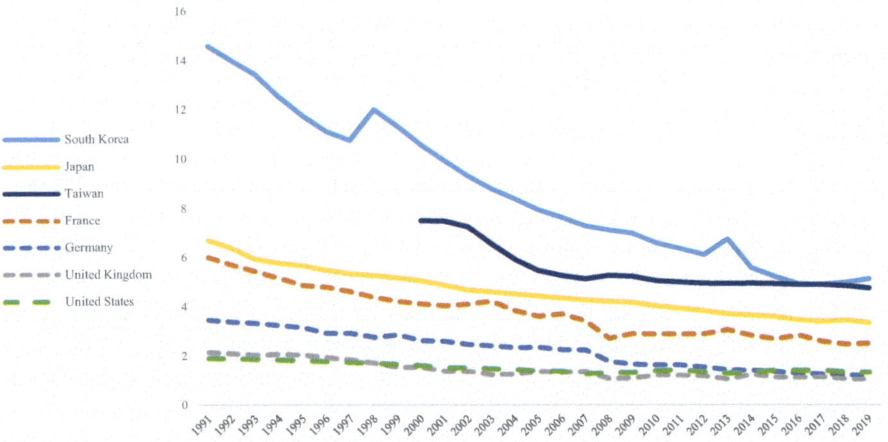

Fig. 1 Employment in agriculture (% of total employment). *Source* (except Taiwan): International Labour Organization, ILOSTAT database. Data were retrieved on January 29, 2021; *Source* (Taiwan): Manpower Survey, National Statistics of the Directorate-General of Budget, Republic of China (Taiwan)

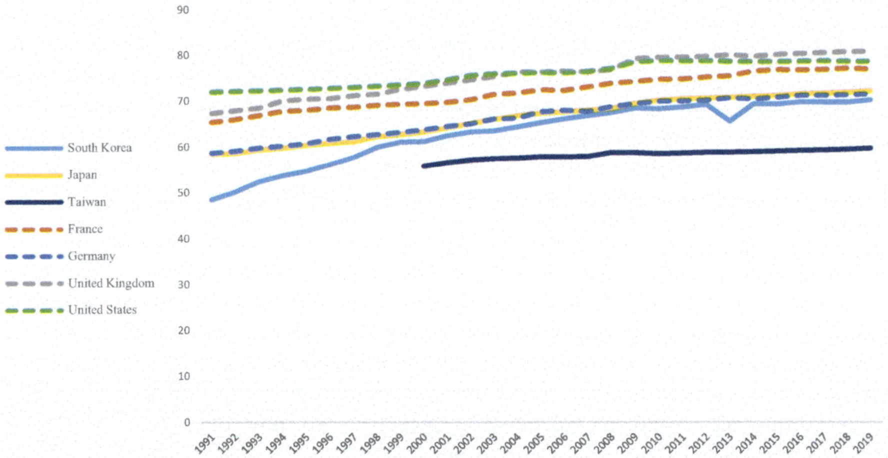

Fig. 2 Employment in services (% of total employment). *Source* (except Taiwan): International Labour Organization, ILOSTAT database. Data were retrieved on January 29, 2021; *Source* (Taiwan): Manpower Survey, National Statistics of the Directorate-General of Budget, Republic of China (Taiwan)

2.2 *Women's Careers and Employment in East Asian Countries*

There are several possible differences in the characteristics of women's employment across East Asia. C First, comparing women in East Asia, they differ in whether their life course forces them to exit the labor market. And they differ in the impact of career interruptions on status attainment. For example, in Japan, women often leave the workforce after marriage or childbirth to become full-time homemakers, which causes them to interrupt their careers. Furthermore, when women re-enter the labor market, they are often hired part-time with low wages, except for some specialized jobs that require specific qualifications. Consequently, women's rate of participation in the labor force in Japan declines from their late twenties to their early 30s and then rises when they re-enter the labor market as part-time workers in their late 30s.

Such a transition in the participation rate of women in the labor force is called "M-shaped employment" (Ochiai 2013), which is also observed in South Korea today (Fig. 3) and prevailed in Taiwan in the 1990s (Brinton 2001). In Taiwan, the female labor force participation rate currently does not increase after the age of 30, but peaks at the age of 20 and thereafter continues to decline (Fig. 3). M-shaped employment shows that it is difficult for East Asian women to balance marriage and childbirth with career continuity. In Japan, the gender segregation of occupations tends to become clearer when women who join professional occupations early in their careers leave them in the middle of their careers (Ikeda 2021). The U.S., a typical industrialized society, demonstrates that the gender segregation of professional and managerial positions has been shrinking faster than that of blue-collar positions since the 1970s (Blau et al. 2013). The difficulty of career continuity makes it difficult for East Asian women to maintain careers even if they are highly educated. Moreover, since the accumulation of work experience is important for managerial positions, it is inferred that it is difficult for East Asian women to obtain such positions.

Additionally, even if East Asian women enter the labor market as full-time workers, if their careers are interrupted, they are forced to work under unstable employment conditions when they re-enter the labor market. This is the second characteristic of women's employment in East Asia. In Japan, the probability of finding regular employment does not decline for young men who experience career interruptions, but for women, career interruptions reduce the probability of finding regular employment, in all cases (Mugiyama 2017).

The fact that East Asian women tend to have to interrupt their careers after marriage or childbirth is often explained by the relationship between family and social institutions. In the welfare state model, familialism means that the family should take maximum responsibility for the welfare of its members, both in income distribution and care sharing (Esping-Andersen 2009). In countries with familialism regimes, women are expected to engage in domestic work and be peripheral when they enter the labor market. Chang argues that the two stages of modernization experienced by Western countries continue to be unbroken in South Korea, as rapid industrial structural changes, globalization, and neoliberal employment system changes occur simultaneously. Moreover, Chang describes this phenomenon as "compressed modernity." According to Chang, in a society undergoing compressed modernity, social institutions that do not conform to the times do not protect the young, and only the family becomes a safety net. Ochiai argues that the promotion of family-oriented policies, as Chang (2010) argues, is also observed in Japan, where the Japanese government has systematically adjusted women's working hours and forced women

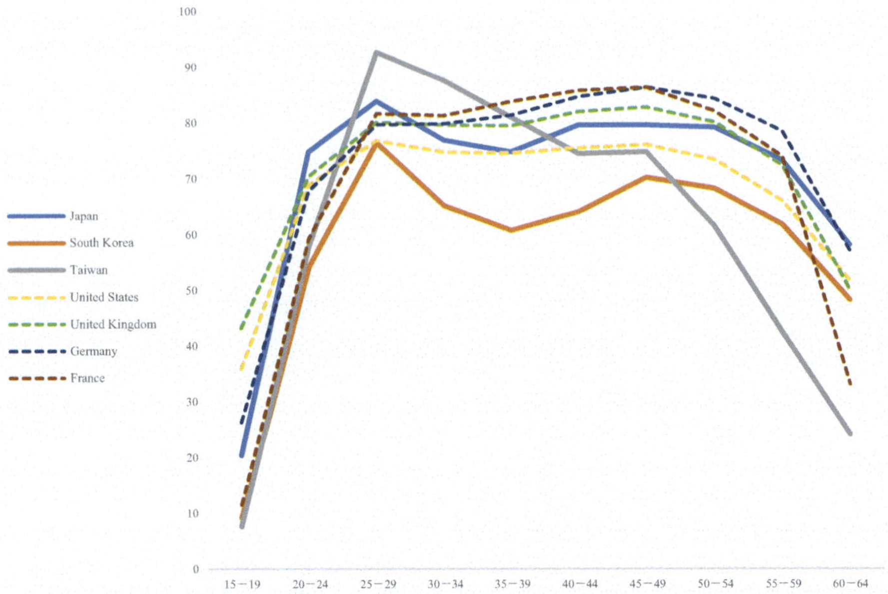

Fig. 3 Female labor force participation rate by age group in 2019. *Source* (Taiwan): Manpower Survey, National Statistics of the Directorate-General of Budget, Republic of China (Taiwan); *Source* (Japan): Labour Force Survey, Statistics Bureau of Japan; *Source* (Other Areas): OECD Database. *Note* (Japan) Domestic residents are included, excluding foreign diplomats and U.S. military residents. Members of the Self-Defense Forces and inmates of confinement facilities are included. The actual number in thousands is not released, so all figures are set to zero; *Note* (the U.S., the UK, Italy) Aggregation is based on 16 years and older

to engage in care work, such as caring for the elderly, to confine women to domestic work to maintain the family unit. Ochiai (2013) points out that these institutional barriers have formed the M-shaped employment pattern of women in Japan.

However, when focusing on care work, Taiwan differs from Japan and South Korea; Ochiai and Monoly (2008) point out that Taiwanese families tend to hire foreign workers for domestic work and elderly care work. They argue that Taiwanese families do not disregard elder care as family work, but consider it as something meaningful for the family to provide, even if it is purchased in the market. As presented in Fig. 3, the participation rate of women in the labor force is higher in Taiwan than in Japan and South Korea, and differs from the M-shape. Thus, Taiwan's approach differs regarding family and women's employment in East Asia. Overall, East Asian countries, other than Taiwan, differ from the West.

2.3 *Differences in Education and Status Attainment: The Circumstances of Women in East Asia*

Comparisons across East Asia show differences not only in women's employment but also in women's educational and status attainment. Brinton (2001) conducted an East Asian comparison of gender segregation in Japan, South Korea, and Taiwan, which are late-industrialized capitalist societies. As Brinton argues, there is a clear gender gap in employment status in Japan and South Korea. Moreover, Brinton contends that the gender gap tends to be clearly visible in the type of employment in Japan and South Korea. The situation in Taiwan differs from that in Japan and South Korea, because many women are self-employed. According to Brinton and Lee (2001), higher education for women continues to expand in Taiwan and South Korea, but not as much in Japan. In addition, Brinton (2001) points out that the expansion of education for women in Taiwan has affected women's employment, but this is not the case in South Korea. Furthermore, Brinton (2001) asserts that in Japan, most women work as non-regular employees, while in Taiwan and South Korea, many women are self-employed or family employees. These differences among East Asian countries relate to variances in the gender gap in wages (Chang and England 2011).

The structure of women's status attainment has also been found to differ in East Asia. International comparisons of social stratification in Japan, South Korea, and Taiwan were conducted by members of the Social Stratification and Social Mobility Survey (SSM 2005). For example, Nakao (2011), who is a main member of the SSM 2005 and who estimated Blau and Duncan's (1967) model of status attainment in three East Asian countries using SSM 2005 data, found a direct effect of fathers' education on incumbency only for Japanese women. Arita (2011) compared the factors that determine wages and subjective self-esteem in three East Asian countries: Taiwan's results demonstrated that occupation impacted wages and subjective self-esteem greatly, and the difference between men and women is small. In South Korea, the effect of occupation and education on subjective self-esteem is strong. In Japan, employment conditions and firm size have an effect larger than in Taiwan and South Korea. Sato (2011) reviewed previous studies and found that the existence or non-existence of the effect of human capital in the form of education is related to the gender gap in the labor market in Japan and South Korea.

3 Analytical Framework and Method

We examined our analytical framework based on the similarities and differences in East Asia, as described above. First, we find that industry and occupational structures are common. Therefore, if the distance between occupations is determined by the commonality of occupational distribution in East Asia, the results of the analysis in South Korea and Taiwan will resemble those of Japan in chapter "Multidimensional Structure of Occupational Segregation by Education and Gender: Time Series Comparison in Japan from 1980 to 2010". In this case, occupational segregation by gender rather than by educational background is observed in South Korea and Taiwan, as in Japan. Furthermore, a commonality between Japan and South Korea but not with Taiwan is the relationship between women's careers and employment. If occupational segregation is clearly indicated by the type of employment and career, we can expect to obtain the same results in Japan and South Korea, but different ones in Taiwan. Finally, if women's higher education and status attainment processes affect the distance between occupations, we may obtain different results for each country. If the lack of expansion of higher education and the intense hierarchical closure among Japanese women create segregation, we should find that Japan has the clearest gendered occupational segregation in East Asia. We add employment status and class of origin to the variables used in chapter "Multidimensional Structure of Occupational Segregation by Education and Gender: Time Series Comparison in Japan from 1980 to 2010 ", namely occupation, gender, education, and age. In this chapter, we interpret age as a proxy for career.

We take a multidimensional view of the distance between occupations with educational experience, gender, and age. Therefore, we adopt a different method from the indexing, which is often used in gender segregation studies. In the study of gender segregation, gender-specific occupational composition and sex ratio in occupations are required, and appropriate and precise indexing of these has been promoted (Charles and Grusky 2004).[1] Although these indexes are appropriate to compare the degree of occupational segregation by gender, they are not suitable to obtain an overall perspective that considers multidimensional factors. For example, the various indicators of occupational segregation by gender, described above, can be defined as a combination of two variables, gender, and occupation, as in the case of *D* introduced earlier. However, if we consider education and age groups separately, the number of indicators for separation becomes large. If numerous indices are compared between the three countries, the indices calculated to aid interpretation will complicate it.

We adopt correspondence analysis as the analytical method. Correspondence analysis, such as *D* and other indexing methods, uses contingency tables, but both the total number of rows and the total number of columns affect the overall analysis results. In other words, this is a "margin-free" method. Therefore, correspondence analysis is also effective in studies on gender segregation. Additionally, correspondence analysis is not only margin-free but also a non-linear method that can visualize various factors surrounding gender segregation. Furthermore, it is easy to add variables to the plot by further dividing the row and column profiles by arbitrary variables. In other words, it is possible to visually capture the separation between individual occupations and gender, considering factors such as education, age, type of employment, employment status, and class origin. Our research began with an interest in understanding the distance between occupations from multiple dimensions, and correspondence analysis enables evaluation based on such an interest (see also chapter "Multidimensional Structure of Occupational Segregation by Education and Gender: Time Series Comparison in Japan from 1980 to 2010"). Therefore, correspondence analysis is appropriate in this study.

4 Data and Variable

4.1 Data

In this chapter, unlike in other chapters using census data, we use survey data that are comparable across East Asia. We needed data that would consider employment status and allow comparisons of occupation and education across East Asia. The availability of the SSM survey and the East Asian Social Survey (EASS) has been a major factor in making international comparisons possible and with ease. For studies of occupation and class mobility using census data, it is difficult to match the unique occupation codes attached to the available data in Japan and South Korea; thus, comparisons are not easy. By categorizing the data by the International Standard Classification of Occupations (ISCO) codes based on free descriptions, the EASS data made it possible to capture East Asia on the same scale.

The EASS data are a bundle of internationally comparable data from four surveys conducted in China, Japan, South Korea, and Taiwan, with questions designed to be comparable. The data used in this study is a merged version of the EASS-2008, 2010, and 2012. The data used merged three periods, EASS-2008, 2010, and 2012, because it was necessary to secure the sample size. Please refer to Table 1 for detailed information on the survey, including the subject of research, sampling method, number of respondents, and response rate. The details of the data are presented in Table 1.

4.2 Variables

In this chapter, we adopt the occupational categories of the ISCO-88 major group for all countries; we exclude armed forces from the comparison because of the under-sample in some regions. For educational qualifications (educational background), the number of years of education up to the last school graduation was used as a category. Due to the size of the sample, 9 and 12 years of education were merged. And we used samples whose age was limited to 20–59 years.

5 Analysis

5.1 Descriptive Statistics

First, we reviewed the proportion of men in various occupations in three countries (Table 2). In all three, the percentage of males employed as managers, technicians, skilled agricultural practitioners, and craft, plant, and machine operators exceeded 50%. In addition to managers and technicians, with high skill levels, the proportion

Table 1 Details of the survey data used in the analysis

	Japan	South Korea	Taiwan
Subject of research	Men and women 20–89 years old	Men and women over 18 years old	Men and women over 19 years old
Extraction method	Stratified two-stage random sampling	Stratified three-stage random sampling	Stratified three-stage random sampling
Planned specimen	4,003 (2008)	2,500 (2008)	4,601 (2008)
	4,500 (2010)	2,500 (2010)	4,424 (2010)
	4,500 (2012)	2,500 (2012)	4,104 (2012)
Title	Japanese general social surveys (JGSS)	Korean general social survey (KGSS)	Taiwan social change survey (TSCS)
Period of implementation	October–December 2008	June–August 2008	July–September 2008
	February–April 2010	June–August 2010	July–November 2011 (February–April 2012)
	February–April 2012	June–August 2012	July–October 2012
Research method	Interview and detention methods combined	Interview method	Interview method
Number of valid responses	2,160 (2006)	1,508 (2008)	2,067 (2006)
	2,496 (2010)	1,576 (2010)	2,199 (2010)
	2,335 (2012)	1,396 (2012)	2,134 (2012)
Recovery rate	60.6% (2006)	61.0% (2008)	44.9% (2006)
	62.1% (2010)	63.0% (2010)	49.7% (2010)
	58.8% (2012)	55.8% (2012)	52.0% (2012)

Table 2 Percentage of students who entered university

	Male		Female	
	On the job (%)	All samples (%)	On the job (%)	All samples (%)
Japan	35.3	33.5	17.0	14.7
Soth Korea	35.7	35.1	26.5	23.5
Taiwan	29.1	28.4	30.5	24.6

of males is high in manual occupations, while the ratio of females is high among clerks, services, and elementary occupations, exceeding 50% in all three countries. The ratio of males to females was approximately the same in all three countries. In the case of professionals, the proportions of male and female workers were similar. The high proportion of male workers in managerial positions is common to all three countries: 88.9% in Japan and 85.7% in South Korea, which is more than 10% higher than that in Taiwan (73.5%) (Fig. 4).

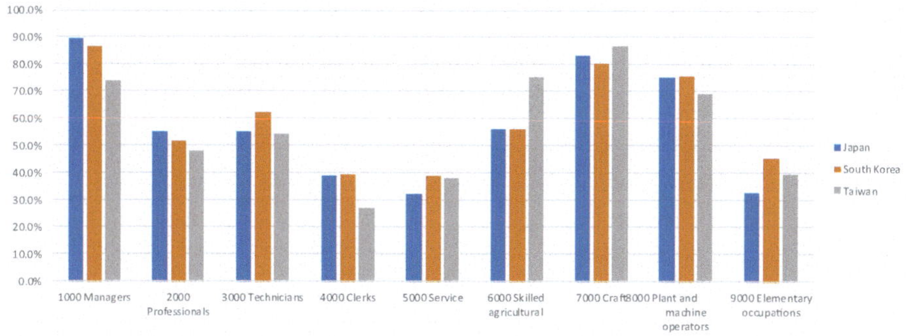

Fig. 4 Percentage of males in each occupation

Then, is it because East Asian women, especially in Japan and South Korea, lack higher education that they have difficulty obtaining managerial positions? In Table 3, we calculate the percentage of employed people who have graduated from university by gender. Table 3 shows that the percentage of male workers with a college degree is between 29 and 35% in all three countries. However, the percentage of women in the workforce with a college degree is much lower in Japan, at only 17%. In Japan, the percentage of university graduates was calculated for the entire sample, including those who were unemployed, because many women who graduated from university left the workforce after marriage or childbirth. Consequently, Japan has the lowest percentage of female college graduates in East Asia (14.7%). On the other hand, the number of female college graduates in South Korea is not significantly different from that in Taiwan in the total sample including the unemployed. In other words, Japan has fewer women with higher education, but South Korea does not have far fewer women with higher education than Taiwan does.

Are the occupations selected by university-educated women different from those of university-educated men? This shows the various occupations among university graduates. In East Asia, the ratio of female university graduates in managerial positions is remarkably low. The total percentage of females in 1000–3000 occupation types is 48.8% in Japan, which is clearly lower than that of South Korea (61.4%) and Taiwan (57.5%). Therefore, even within East Asia, Japan and South Korea have a low percentage of women in management positions among university graduates. In addition, Japan has a low percentage of women with college degrees in the 1000–3000 occupation types, which are occupations with higher skill levels. Among the East Asian countries, the gender gap in higher education is diverse, but there also seem to be differences in careers after graduation.

To summarize the results so far, East Asia has a common tendency of vertical segregation among the positions with high skill levels, as the percentage of women in the manager and technician positions is low. In East Asia, the proportion of women in manual positions is also low, indicating a tendency toward horizontal segregation. In Japan, the number of women who graduate from university is lower than in other East Asian countries, and the number of university graduates in managerial positions

Table 3 Occupations of people with higher education

	Japan		South Korea		Taiwan	
	Male	Female	Male	Female	Male	Female
1000 Manager	11.6%	0.6%	8.0%	1.4%	13.6%	5.9%
2000 Professionals	24.9%	30.5%	18.6%	27.1%	23.2%	22.2%
3000 Technicians	18.0%	17.7%	41.0%	32.9%	31.6%	29.4%
4000 Clerks	23.4%	33.5%	13.5%	24.7%	9.7%	29.4%
5000 Service	7.8%	13.4%	9.8%	11.2%	7.1%	10.4%
6000 Skilled agricultural	1.0%	0.6%	1.4%	0.3%	0.6%	0.5%
7000 Craft	4.9%	0.9%	2.9%	0.3%	5.8%	0.5%
8000 Plant and machine operators	5.4%	0.9%	2.5%	0.3%	5.0%	0.7%
9000 Elementary occupations	3.0%	1.8%	2.3%	1.7%	3.2%	0.9%
Total percentage of 1000s to 3000s	54.5%	48.8%	67.6%	61.4%	68.4%	57.5%
N	795	1596	4178	880	817	295

is low in Japan and South Korea. In Japan, the number of women who graduate from universities is low compared to other East Asian countries. Thus, even in East Asia, it is difficult for women in Japan to receive education and find high-skill jobs even after graduating from university.

5.2 *Basic Correspondence Analysis: Occupation, Education, Gender, and Age*

Based on a comparison of these proportions, we confirmed the results of the correspondence analysis.[2] Figure 5 shows the eigenvalues and cumulative contribution rates for each dimension analyzed using correspondence analysis. For all countries, the changes in both the eigenvalues and cumulative contribution rates were small after Dim 2. This is called an "elbow" in the results of correspondence analysis and is generally interpreted by projecting the dimension up to the point where the change becomes gradually expressed in space. Therefore, the number of dimensions to be projected in space was set to two. Taiwan and South Korea have the largest Dim 1, with a contribution rate of 54.7%, and Japan has the smallest Dim 1, with a contribution rate of 51.2%. The largest Dim 2 is in Japan, with a contribution rate of 29.6%. The smallest contribution rate was 19.9% in South Korea. In other words, the characteristics of Dim 1 are strongest in Taiwan and South Korea and Dim 2 is the strongest in Japan.

We interpret the two main dimensions that were analyzed and their distribution in each category. In Fig. 5 we plot the coordinate points for each category, analyzed by correspondence analysis. In this plot, the left column shows the coordinate points of sex, education, and gender for each dimension, and the right column shows those for

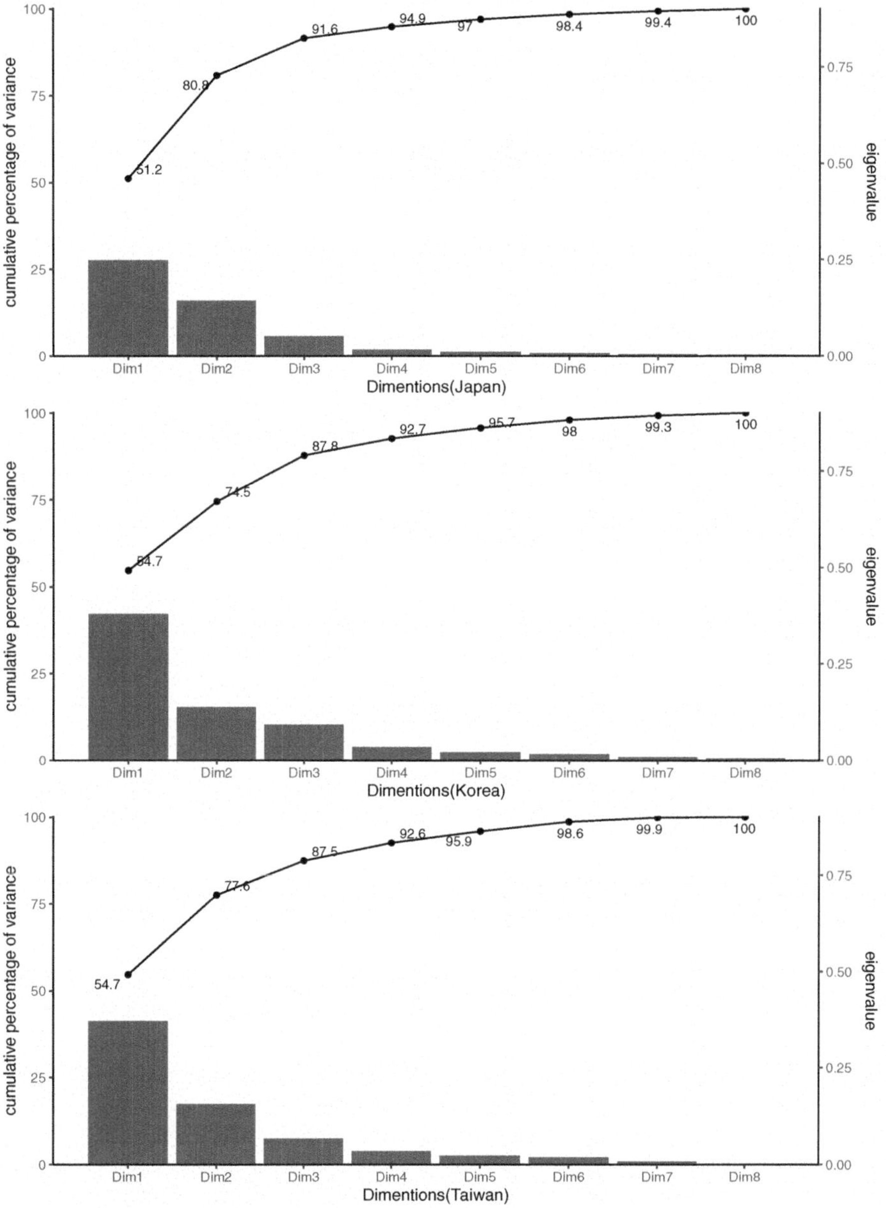

Fig. 5 Eigenvalues and cumulative percentage of variance

occupation (ISCO-88 major group) for each dimension. For both Dim 1 and Dim 2, there are common coordinate characteristics among the three East Asian countries. Dim 1 tends to be distributed in a positive direction with low education (12 years or less education when Dim 1 is positive) and in a negative direction with high education (16 years or more education when Dim 1 is negative) in all countries. In the plot of occupations in the right column, the number of occupation categories (1000–9000) corresponding to the skill level increases from left to right. In other words, occupations with high skill levels tended to be distributed in the negative direction, while occupations with low skill levels tended to be distributed in the positive direction. Therefore, Dim 1 can be interpreted as an axis of education and occupational skill level. On the other hand, Dim 2 tends to have males distributed in the positive direction and females in the negative direction, bordering on the origin. Therefore, Dim 2 can be interpreted as an axis of gender. Thus, when we conduct a correspondence analysis of East Asia by sex, education, and age using the ISCO-88 classification, we can analyze the education/skill level axis and the gender axis with a common basis. Given the contribution rates (Fig. 5), more than 70% of the variance can be explained by the variance on these axes.

What are the characteristics that differ among East Asian countries? First, we focus on the fact that the percentage of contribution of each dimension differs across countries, and the highest contribution rate of Dim 1 is found in Taiwan, suggesting that a relatively large portion of the occupation-gender-age contingency table in Taiwan can be explained by education and skill level. Alternatively, the contribution percentage of Dim 2 is the highest in Japan, suggesting that gender has a relatively large part in explaining the cross table. South Korea's results are positioned in the middle of the other two countries. These characteristics indicate that occupational segregation by gender, age, and education is relatively widespread in Japan in East Asia.

Thereafter, we examine the individual plots to interpret the characteristics of occupational segregation based on gender, in each country. In this analysis, we use the ISCO code, and for the sake of interpretation, we briefly explain the classification criteria of ISCO-88. Skill level in ISCO-88 is a classification criterion for occupations based on the complexity of tasks and duties, but it is coded to correspond to the level of education. Therefore, it is expected that an education level of 16 years or more will be plotted in correspondence with managerial, professional, and technical occupations.

The plots of occupations in Japan show that university graduates in their 30s–50s are in the same quadrant as managers, professionals, and technicians. Dim 2 is a gender axis, which indicates that management positions in the survey data are dominated by older men. Dim 2 is an axis of gender, which indicates that older men occupy managerial positions in the survey data. For men in their 20s with a college degree, Dim 1 is in the negative direction, but so is Dim 2. The positive value of Dim 2 increases as the age group increases for men in their 30s to those in their 50s. The same trend is observed for men with other educational backgrounds, but it is not as clear as for men with university degrees.

The results for Japanese women were interpreted in more detail. In Dim 1, Japanese women are distributed in a negative direction from around the origin. If we simply interpret the results in terms of skill level, we might conclude that women are employed in occupations with relatively high skill levels, but this is not the case. Since the contribution of Dim 2 is relatively high in Japan, it is necessary to interpret the results based on Dim 2. The occupations where Dim 2 is negatively distributed are agriculture, labor, personal services, and clerical work, and in particular, service and labor occupations where the absolute value of Dim 2 is large often have unstable employment and wages that are not high. This is where less educated women are concentrated. What about the case in which women acquire education? As the age of college-educated women increases, Dim 2 shifts in the positive direction. This is the same as that in men. However, even when they rise, they do not stay in the same positions as men, but rather in technical and professional positions.

The plots of gender, education, and age in South Korea show a clearer separation in Dim 2 than in Japan. In Japan, women with 16 or more years of education occupy the second quadrant, as are men, but in South Korea, they occupy the third quadrant. In South Korea, managerial, professional, and technical occupations are located in the second quadrant of the occupational plot. However, since the coordinates of Dim 2 are close to 0 for professional and technical occupations, they are near to the coordinates of university-educated women. The effect of education on the number of managers may be higher than that in Japan because managers are slightly in the positive direction of Dim 1 compared to Japan. In South Korea, the occupations in which Dim 2 is negatively distributed are laborers, personal service workers, and office workers. In the gender-education-age plot, less-educated women are concentrated in coordinates similar to those of service and labor occupations. This tendency is common in Japan. However, males in their 20s with 12 years of education or fewer are characterized by coordinates corresponding to service and labor occupations.

Finally, we reviewed the results for Taiwan. The characteristics of the plot for gender, education, and age in Taiwan are that the segregation structure is clear through education. Both men and women in their twenties with 16 or more years of education approximate the 14-year education category. The occupation plots did not differ significantly from those of Japan and South Korea.

5.3 Correspondence Analysis Considering Employment Status and Class Structure

Next, we confirm the segregation structure based not only on the occupational area but also on employment type. Figures 7, 8, and 9 show the results of the correspondence analysis when the ISCO-88 occupational categories are classified by employment type. The classification method was as follows: First, employers are not classified by job type, but are treated as a single category. Managers and skilled agricultural workers were treated as a single category. The other occupations were classified into four categories: regular workers, temporary workers, self-employment, and family workers. In the legend, shapes identify employers, managers, and skilled agricultural workers, colors identify status, and numbers are ISCO-88 codes for major occupational categories. The numbers are ISCO-88 occupation codes.

Compared to Figs. 6 and 7 shows a clear separation structure in the gender, education, and age plots; men and women are separated by a 45-degree line dividing Dim 1 and Dim 2. In the occupation plot, women take the coordinates corresponding to temporary workers (and men in their 20s are included). The coordinates of managers and employees are located near the zero of Dim 1 in Fig. 6. But in Fig. 7, they are positioned in the negative direction of Dim 1. This is between college-educated men and non-college-educated men. In other words, dividing occupations by employment status and analyzing employers as independent categories makes the relationship between managers or employers and education less clear, while it tends to be more defined by gender.

Figures 8 and 9, the plots for Korea and Taiwan, do not appear to differ from Fig. 6. In other words, the impact of employment status on the plots in Korea and Taiwan is not so large, unlike in Japan. Although previous studies indicate that there are many women who are engaged in self-employment in South Korea and Taiwan, there is no tendency in the plots of gender, education, and age that the attributes of the positions close to the coordinates of self-employment are all women. In addition, unlike in Japan, temporary workers are also found in the first quadrant, and the position of women in the gender-education-age plot in South Korea and Taiwan cannot be interpreted as temporary workers in the occupation plot. The coordinates of Dim 1 are close to those in Fig. 6 for managerial positions in the plots of occupations in South Korea and Taiwan.

Finally, we show the results of partitioning gender, education, and age by origin class (Fig. 10). The origin hierarchy was manipulated using the father's education (binary of 16 years or more, or 14 years or less).

Looking at the results for Japan (Fig. 10), there is no significant change in the occupation plot compared to Fig. 6. In the gender-education-age plot, Dim 2 tended to be positive when fathers' education was high for females. In the case of women, Dim 2 tends to be positive if their fathers are highly educated. When the father is a college graduate and the child is a college graduate, the coordinates of Dim 2 are close to those of the male college graduate. This is a remarkable difference compared to that in Fig. 6.

In South Korea (Fig. 11), the effect of the class of origin on occupational segregation is not as clear as in Japan for both men and women compared to Fig. 6. However, in Taiwan (Fig. 12), the influence of one's class of origin on occupational segregation appears to be present in the plot. In the gender-education-age plot, Dim 2 tends to be positive for both college-educated men and college-educated women whose fathers are college graduates. In other words, we can confirm that both males and females are separated by their class of origin. In Japan, the separation of occupational categories by class of origin is more pronounced for women, but in Taiwan, the separation of occupational categories by class of origin can be confirmed for both men and women.

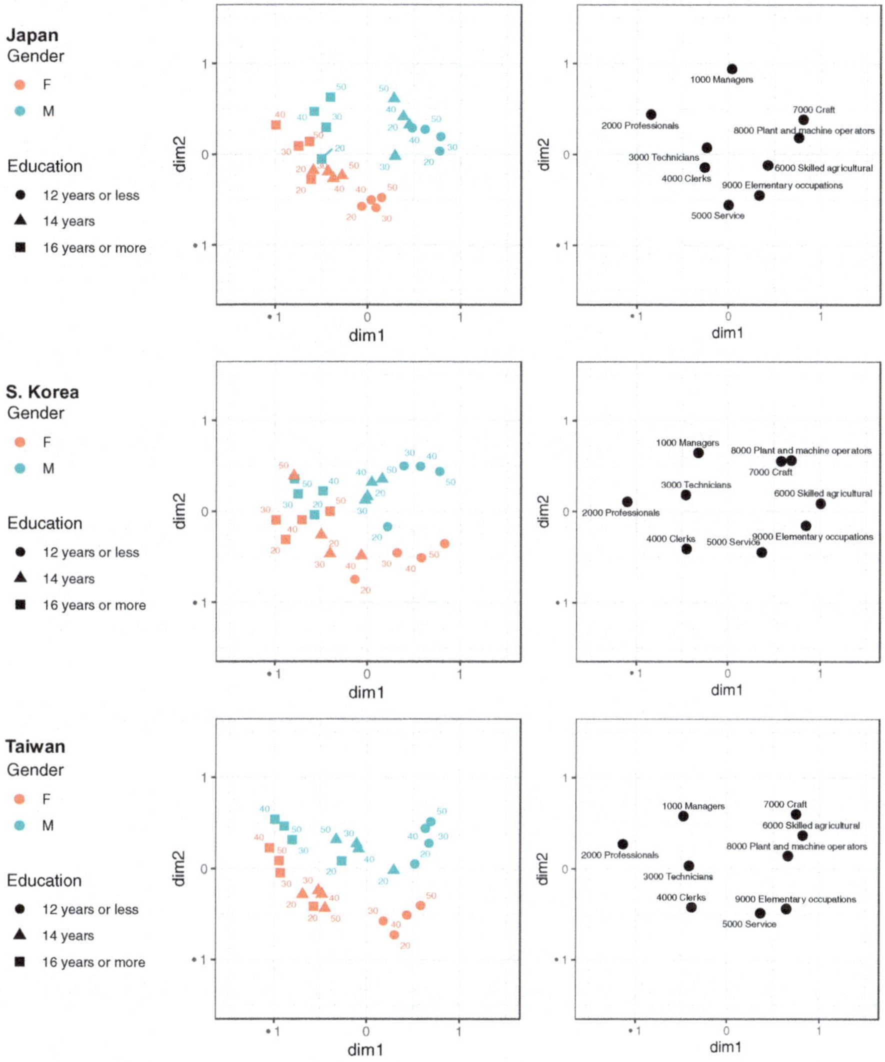

Fig. 6 Results of correspondence analysis (gender, age, education, occupation)

6 Conclusion

As a result of the analysis, the following points became clear. The comparison of occupational structure by frequency distribution shows that there are some commonalities among East Asian countries. East Asia shares a high ratio of male workers in managerial and manual positions. In addition, we conducted a correspondence analysis that allowed us to identify a space consisting of two dimensions that are common in East Asia. The first dimension of the space, Dim 1, is the axis related to education

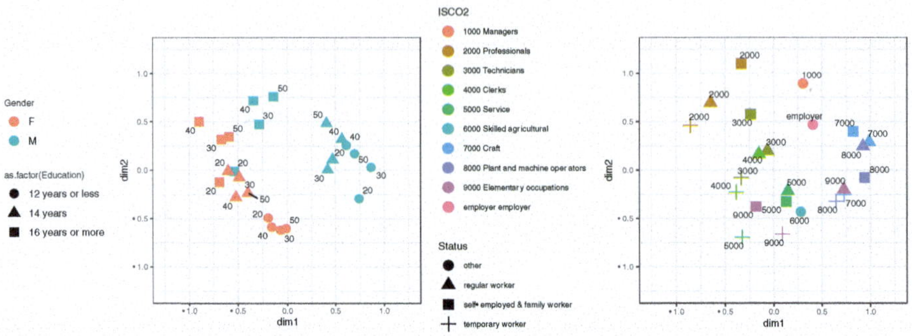

Fig. 7 Results of correspondence analysis (employment status, Japan)

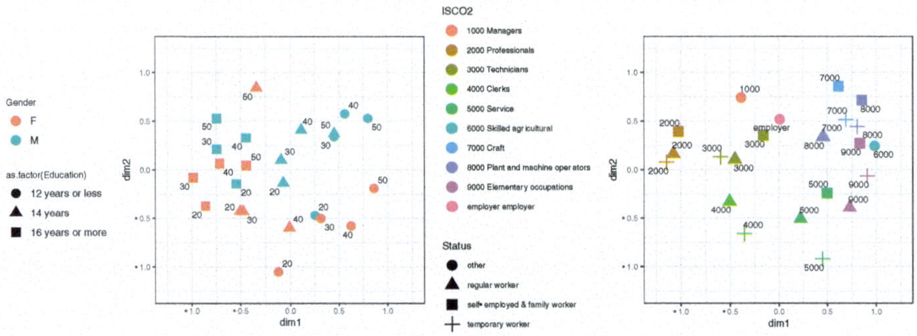

Fig. 8 Results of correspondence analysis (employment status, South Korea)

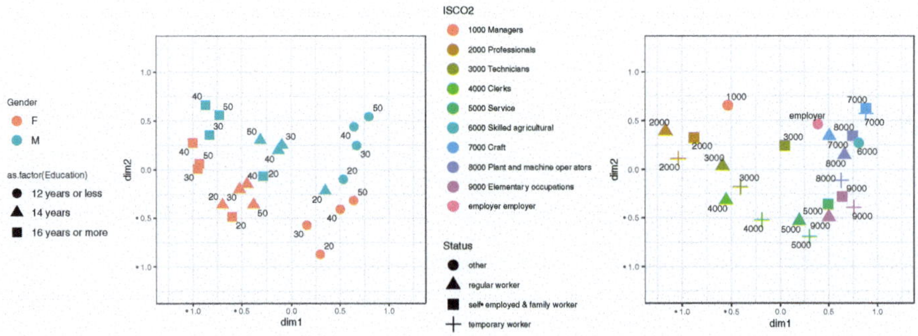

Fig. 9 Results of correspondence analysis (employment status, Taiwan)

and occupational skill levels for all countries, including Japan. This characteristic of Dim 1 is understandable because the skill level defined by ISCO includes criteria based on educational qualifications. In East Asia countries, the positions of college graduates in the gender, education, and age plots correspond to the positions of those with high skill levels in their occupations. And, it is also common for women in East

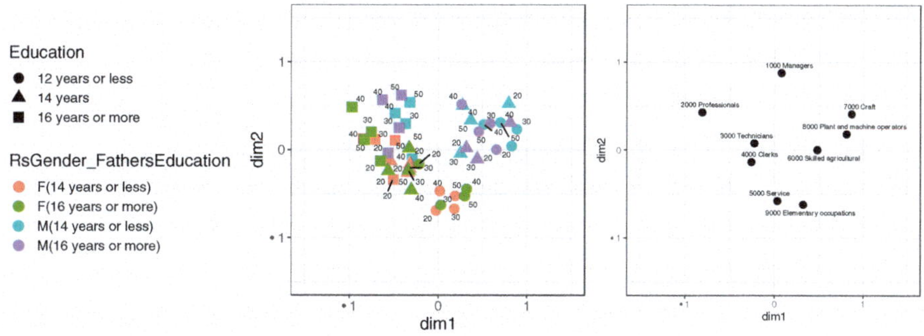

Fig. 10 Results of correspondence analysis (gender, age, education, occupation, and class structure, Japan)

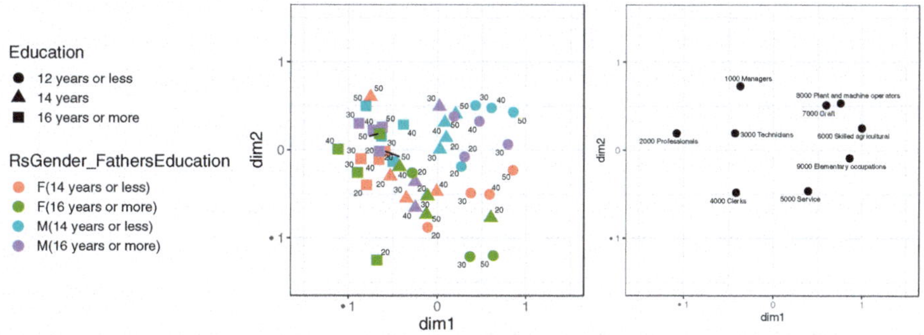

Fig. 11 Results of correspondence analysis (gender, age, education, occupation, and class structure, South Korea)

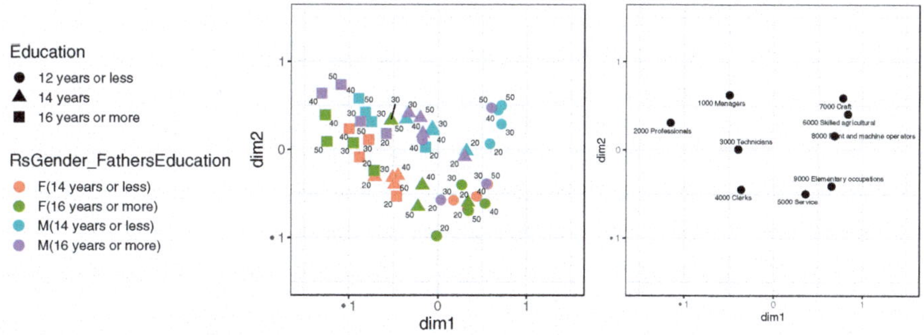

Fig. 12 Results of correspondence analysis (gender, age, education, occupation, and class structure, Taiwan)

Asia to hold managerial positions. This implies that gender asymmetry in education is common in East Asian societies.

The differences in the East Asian comparison are in the contribution of both dimensions, and in the position of women in the gender-education-age plot. Dim 2 in Fig. 6 can be interpreted as a gender axis, and the country with the highest contribution of Dim 2 is Japan. Therefore, in the occupational segregation in East Asia, gender-based segregation is more pronounced in Japan. Furthermore, when we subdivided the occupational categories according to employment status, the plots for gender, education, and age in Japan changed compared to the plots without employment status. If we focus on the position of each category, we can observe a clear occupational separation. This result is not observed in South Korea or Taiwan. In particular, the position of women in the gender-education-age plot corresponds to the position of temporary workers in the occupation, which is characteristic of Japan. Furthermore, when we check the effect of occupational segregation by class of origin, we find that in Japan, the plot for women with a university degree resembles that for men with one. Similarly, in Taiwan, the plots for gender, education, and age change depending on one's class of origin, which change is independent of gender.

These results differ from the assumption in some respects. We expected that gender segregation by employment status would be similar in Japan and South Korea; however, segregation by non-regular and regular employment is clear in Japan and not in South Korea. The career path of women may be different in Japan and South Korea, where the transition of M-shape-type employment is similar. In Japan, women leave their first job with a high skill level and re-enter the workforce as non-regular workers. In South Korea, there is a relatively strong segregation of occupational categories, so women may be disadvantaged in their first job.

In addition, we expected that occupational segregation by class of origin would be similar in Japan and Taiwan, but in Japan, its effect is felt mainly by women, while in Taiwan, the effect is found both in men and women. In this sense, the Taiwanese hierarchy may be more closed than Japan, but this may be due to the relatively few college graduates. In other words, there are few college graduates among males and females, but the influence of their class of origin on obtaining a college degree may be significant.[3] On the other hand, men tend to be less affected by their class of origin, in Japan, which may be because there are more male than female college graduates.

In East Asian societies, which share some similarities in the industrialization process, gender-based segregation of occupations exists, with manual labor and managerial positions dominated by men. Moreover, in East Asia, Japan is characterized by unstable employment of women due to their exit from and re-entry into the labor market, and their rate of participation in higher education, which is strongly influenced by their class of origin. This is a significant difference among East Asian countries.

In this paper, we examined the similarities and heterogeneity of gender occupational segregation in East Asia, which is often categorized as having the same social type due to similarities in industrial structure, employment, and family characteristics. In this process, the characteristics of Japan have gradually become clear. However, further analyses are required. That is an international comparison that includes Europe and other Asian countries. The inclusion of these countries, as well as the U.S. and East Asian countries, would help clarify Japan's position among countries with different degrees of progress in industrialization.

Notes

1. For example, we examine the "index of dissimilarity" (D), which is the most popular index in the study of occupational gender segregation.

$$D = \sum_{j=1}^{J} \left| \frac{F_j}{F} - \frac{M_j}{M} \right| \times 100 \times \frac{1}{2}$$

   In the above formulas, j refers to each occupation; F_j or M_j is the total number of women or men with an occupation j, F or M is the total number of women or men, and D is the total percentage of the male or female labor force that must be subtracted compared to the case where the ratio of men to women in occupation j is perfectly equal. Charles and Grusky, using the formula for D as an example, posited that a major problem with indexing is that it is "margin-dependent." The term "margin-dependent" refers to a situation in which the degree of gender segregation depends on the occupation side or gender side when the degree of gender segregation is indexed from the occupation and gender contingency table. Here, the number of male or female workers in an occupation does not affect the size of D. In other words, it is impossible to evaluate the bias of the sex ratio in an occupation or in the entire labor market. Later, indices that control the sex ratio (Size-Standardized Index: Ds) and that evaluate both margins (Association Index: A) were developed (Charles and Grusky 2004).
2. In this chapter, we use the CA function in the R package, and FactoMineR, for the calculation. For plots, we use ggplot2 instead of the output function included in the function.
3. However, since the university enrollment rate in Taiwan is rising rapidly, the results may change if the analysis is conducted using data as of 2021.

Acknowledgements This work was also supported by JSPS KAKENHI Grant Number 19J10029, 22J00250.

References

Arita, Shin. 2009. A comparative study of social stratification in East Asian societies. *Japanese Sociological Review* 59 (4): 663–681 (in Japanese).

Arita, Shin. 2011. Higashi-Aija No Syakaikaisoukouzouhikaku (Comparison of social stratification in East Asia). In *Structure of inequality and mobility: Contemporary stratified society 2*, ed. Hiroshi Ishida, Hiroyuki Kondou, and Keiko Nakao, 273–287. University of Tokyo Press (in Japanese).

Blau, Peter M., and Otis D. Duncan. 1967. *The American occupational structure*. New York: Wiley.

Blau, Francine D., Peter Brummund, and Albert Yung-Hsu. Liu. 2013. Trends in occupational segregation by gender 1970–2009: Adjusting for the impact of changes in the occupational coding system. *Demography* 50 (2): 471–492.

Brinton, Mary C. 2001. Married women's labor in East Asian economies. In *Women's working lives in East Asia*, ed. Mary C. Brinton, 1–37. Stanford: Stanford University Press.

Brinton, Mary C., and Yean-Ju Lee. 2001. Women's education and the labor markets in Japan and South Korea. In *Women's working lives in East Asia*, ed. Mary Brinton, and Mary C. Brinton, 125–150. Stanford: Stanford University Press.

Chang, Chin-fen, and Paula England. 2011. Gender inequality in earnings in industrialized East Asia. *Social Science Research* 40 (1): 1–14.

Chang, Kyung-Sup. 2010. *South Korea under compressed modernity: Familial political economy in transition*. London: Routledge.

Charles, Maria, and David B. Grusky. 2004. *Occupational ghettos*. Stanford: Stanford University Press.

Esping-Andersen, Gøsta. 2009. *The incomplete revolution: Adapting to women's new right*. Cambridge: Polity Press.

Goldin, Claudia. 1994. The U-shaped female labor force function in economic development and economic history, NBER Working, National Bureau of Economic Research: paper: 4707.

Ikeda, Takehiro. 2021. Gender differences in professional career mobility under occupational gender-segregated workplace structure. Report on the results of a secondary analysis of actual conditions and changes in employment, income, learning, and living conditions using the Japanese panel study of employment dynamics, center for social research and data archives, 41–55. Institute of Social Science University of Tokyo (in Japanese).

Jackson, M., J.H. Goldthorpe, and C. Mills. 2005. Education, employers and class mobility. *Research in Social Stratification and Mobility* 23: 3–33.

Jacobs, Jerry A., and Suet T. Lim. 1992. Trend in occupational and industrial sex segregation in 56 countries, 1960–1980. *Work and Occupations* 19 (4): 450–486.

Mugiyama, Ryota. 2017. Inequality caused by career interruptions: Examining persistent effects upon getting regular employment. *Japanese Sociological Review* 68 (2): 248–264 (in Japanese).

Nakao, Keiko. 2011. Chiitassei Moderu No Higashi-Ajia Kokusaihikaku (An international comparison of East Asian models of status attainment). In *Structure of inequality and mobility: Contemporary stratified society 2*, ed. Hiroshi Ishida, Hiroyuki Kondou, and Keiko Nakao, 289–300. Tokyo: University of Tokyo Press (in Japanese).

Nakao, Keiko. 2018. Jyosyou Kenkyu No Gaiyou to Mokuteki (Introduction: Overview and objectives of the study), *JSPS KAKEN Grant Research Paper, The Future of the Education Society* (Reports Grant number 15K03818), 1–4 (in Japanese).

Ochiai, Emiko, and Barbara Molony. 2008. Asia's new mothers: Crafting gender roles and childcare networks in East and Southeast Asian societies. Kent: Global Oriental.

Ochiai, Emiko. 2013. HIsgashi Ajia No Teisyusseiritu to Kaokusyugi: Hanassyuku-kindai Tositeno Nihon (Low fertility and familialism in East Asia: Compressed and semi-compressed modernities in Japan). In Shinmitsuken to Koukyouken No Saihensei: Ajia Kindai Karano Toi (Transformation of the intimate and the public in Asian modernity), ed. Emiko Ochiai. Kyoto: Kyoto University Press.

Sato, Yoshimichi. 2011. "Shakai-Kaisou Shakai-Idou Chousa Wo Meguru Kokusaihikaku No Konnan To Kanousei" (Difficulties asnd possibilities of international comparisons on social stratification and social mobility surveys: Experiences from the 2005 SSM survey). *Advances in Social Research* 7: 12–17 (in Japanese).

Shirahase, Sawako, and Hiroshi Ishida. 1994. Gender inequality in the Japanese occupational structure. *International Journal of Comparative Sociology* 35 (3): 188–206.

Treiman, D.J. 1970. Industrialization and social stratification. *Sociological Inquiry* 40 (2): 207–234.

Uchikoshi, Fumiya, and Ryota Mugiyama. 2020. "Nihon ni Okeru Seibetsu-shokuiki-bunri No Susei" (Trends in occupational sex segregation in Japan: A decomposition analysis of census data, 1980–2005). *Population Studies* 56: 9–23 (in Japanese).

Wakita, Aya. 2018. Changes in educational premium: A comparison of the education-occupation linear relationship in Japan and the US. *JSPS KAKEN grant research paper: The future of the education society* (Reports Grant Number 15K03818), 5–20 (in Japanese).

An International Comparison of Occupational Segregation by Gender: Through a Comparison of 14 Countries

Keiko Nakao

1 Introduction

When one tries hard to advance to a higher level of education, one is expected to be rewarded with a higher status of occupation subsequently. This is the basis for meritocracy.

As previous chapters noted, the link between education and occupation has traditionally attracted the concern of sociologists in the context of modernization (e.g., Blau and Duncan 1967; Treiman 1970; Charles and Grusky 2004; Jackson et al. 2005; Torche 2011). There are two theoretical explanations for this relationship. The first is the importance of education for occupational status. Based on the proposition that social mobility positively relates to the extent of industrialization, Treiman's (1970) early study focused on the relationship to argue that the influences of education are stronger in more compared to less industrialized societies[1].

In such a society, inequality based on attributes would be denied and competition based on educational attainment would be supported.

The second is the relative decline in the effectiveness of education due to the increase in the number of highly educated people. Jackson et al. (2005) focused on diminishing returns to educational attainment. They noted that the importance of education depends on the changes happening to a society's industrial economic structure and the proportion of highly educated people. In those societies, higher educational attainment did not ensure high-status occupations. Indeed, Japan and the U.S. can be said to have entered a "universal stage" of educational attainment (Trow 1976). It seems that educational attainment no longer guarantees high occupational status in societies with post-industrial economic structures.

K. Nakao (✉)
Tokyo Metropolitan University, 1-1 Minamiosawa, Hachioji-City, Tokyo 192-0397, Japan
e-mail: nakao@tmu.ac.jp

K. Nakao (ed.), *Multidimensional Aspects of Occupational Segregation*,
Behaviormetrics: Quantitative Approaches to Human Behavior 18,
https://doi.org/10.1007/978-981-99-8513-5_7

Given the two perspectives on meritocracy in industrialized societies, the relationship between education and occupation should be examined considering macro factors such as industrial structure. At the very least, we should be aware of the complexity of the modernization process in European and Asian societies. Chang (1999, 2010) proposed the concept of "compressed modernity" and asserted that Western-derived social theories or arguments could not satisfactorily explain modernization in South Korea. One reason is that Asian societies are modernizing much more rapidly, over 50 years, than Western societies, which have developed over 200 years. Chang's (1999, 2010) argument cautions us about the pitfalls of emphasizing the Western perspective on modernization processes. A historical comparison between modernization in European and Asian countries is beyond the scope of this study. However, Chang's study indicates that it is necessary to consider macroeconomic factors such as industrial structure to understand occupational segregation by gender.

Previous chapters in this book have also conducted international comparisons. In Chap. "Multidimensional Structure of Occupational Segregation in Japan and the United States: Comparison of Trends from 1980 to 2010" of this book, Saitoh focused on this relationship and found multiple aspects of occupational segregation through a Japan-the U.S. comparison. Saitoh pointed out distinct differences between Japan and the U.S. regarding gender tracks. This is the difference between the U.S., where there is relatively small occupational segregation between gender in the highly educated group, and Japanese society, where there is consistent occupational segregation by gender. Saitoh's approach is supported by previous research, such as Featherman et al. (1975), who focused on the commonalities rather than the differences among industrialized societies. Kuwana in Chap. "Comparison of Occupational Segregation in East Asia: Analyzing the Structure of Segregation by Educational Background and Gender" also noted the strength of attributes principal in Japanese society from international comparison. Kuwana pointed out that among the three East Asian countries of Japan, Korea, and Taiwan, occupational segregation by gender was relatively pronounced in Japan.

These studies demonstrate that educational attainment and occupational status are complex. However, their analysis is limited to a few comparisons of a few countries. That is a comparison of the U.S.-Japan comparison and a comparison of three East Asian countries. Indeed, their analysis may be indicative of the characteristics of Japanese society. But if we take the attribution principle of post-industrialization as a theoretical perspective, comparisons among many countries, including Japan, are still desirable. Thus, this chapter considers the relationship between educational attainment and occupational status through an international comparison, considering the macro-level changes occurring at this stage of development, such as changes to industrial economic structures.

2 Framework and the Data

The main interest of this book is to examine the link between industrialization and meritocracy in relation to education and occupation. This chapter's framework follows the previous chapters, which have been concerned with the spread of meritocracy and the correction of gender segregation due to post-industrialization. Of course, each country has its own political system, historical context, and other conditions. Here, we would like to emphasize that the purpose of this study is not to describe the situation specific to each country. Rather than attempting to identify the individual conditions in each country, the interest of this study is to provide a global viewpoint on post-industrialization and meritocracy relationships.

Table 1 is a restatement of Chap. "Introduction": Percentage of Workers by Sector for 16 countries, as published by the International Labour Organization (2022). The Services sector represents tertiary sector workers, Industry represents secondary sector workers, and Agriculture represents primary sector workers. Let us assume for the moment that post-industrialization is the development of the tertiary Industry. It is obvious that while the industrial transformation in all countries has been moving toward post-industrialization, each country's position differs in terms of the stage of industrialization. Table 1 clarifies that the progress of post-industrialization varies from country to country.

We are looking at Japan, as already discussed in Chap. "Introduction", post-industrialization is the most advanced among Asian countries at any point in time and is somewhat behind compared to Western countries. How, then, does the segregation of occupational lines in Japan observed in the previous chapters relate to the trend toward post-industrialization? This chapter analyzes the correspondence between education and occupation in the above countries, paying attention to Japan, and attempts to explain it through an exploratory typology of each country. This chapter uses IPUMS data[2]. Among the above 16 countries, 14 are analyzed, excluding South Korea and Taiwan, for which IPUMS data are unavailable. Since there is no data for Japan in IPUMS, the custom-made tabulation results of Japan's Census of Population (Ministry of Internal Affairs and Communications) will be used as data.

Occupation categories are nine (ISCO-88 major groups): legislators and managers, professionals, technicians, clerks, service and sales, agricultural, crafts, plant and machine operators, and elementary occupations.

Note that the data for Japan is not from the IPUMS, and thus its occupational classification differs from that of other countries. As already explained in Chap. "Int roduction", we will categorize the Japan Data on Professionals into three categories: professional with advanced qualification (professional 1), professional and technical with personal service (professional 2), and professional and technical with personal service (professional 3). However, the purpose of this study is not to examine the detailed relationship between occupation and education in each country but rather to identify overall trends through comparisons among countries. Therefore, the fact that

Table 1 Employment by Sector in 2000, 2010, and 2019

	Year 2000			Year 2010			Year 2019		
	Services (%)	Industry (%)	Agriculture (%)	Services (%)	Industry (%)	Agriculture (%)	Services (%)	Industry (%)	Agriculture (%)
Vietnam	22.3	12.4	65.3	29.6	21.7	48.7	35.3	27.4	37.2
India	24.0	16.3	59.6	26.7	21.8	51.5	32.3	25.1	42.6
China	27.5	22.5	50.0	34.6	28.7	36.7	47.2	27.4	25.3
Thailand	32.2	19.0	48.8	41.1	20.6	38.2	45.7	22.8	31.4
Philippines	47.0	15.9	37.1	51.4	15.7	32.9	58.2	19.0	22.8
Malaysia	49.5	32.2	18.4	59.1	27.6	13.3	62.8	27.0	10.3
Poland	50.3	31.1	18.7	56.6	30.3	13.1	58.7	32.2	9.2
Taiwan	53.3	38.7	8.0	57.3	37.3	5.4	58.2	36.7	5.0
Hungary	59.8	33.8	6.5	64.8	30.7	4.5	63.1	32.2	4.7
South Korea	61.3	28.1	10.6	68.4	25.0	6.6	70.4	24.5	5.1
Italy	63.0	31.8	5.2	67.6	28.6	3.8	70.3	25.8	3.9
Japan	63.5	31.4	5.1	70.4	25.6	4.1	72.4	24.2	3.4
France	69.6	26.3	4.1	74.8	22.3	2.9	77.1	20.4	2.5
UK	73.3	25.2	1.5	79.6	19.2	1.2	80.9	18.1	1.0
U.S	73.9	24.4	1.6	78.9	19.6	1.4	78.8	19.9	1.4
Netherlands	75.1	21.6	3.3	79.2	17.7	3.1	81.8	16.1	2.1

occupational classifications differ in the Japanese data is unlikely to be a significant issue.

3 The Analysis

The analysis of each country is presented in two figures: the first is the composition of occupations, and the second is the attributes of the subjects. As in the other chapters, we will examine the correspondence between attributes and achievements in an exploratory manner through correspondence analysis. Specifically, we will rely on the percentages of Employment by Sector shown in Table 1.

4 Group 1…The Netherlands, the U.S., the U.K., and France

The 14 countries were classified into three groups in terms of post-industrialization. First, let's look at the Netherlands (Figs. 1 and 2), the U.S. (Figs. 3 and 4), the U.K. (Figs. 5 and 6), and France (Figs. 7 and 8). These four countries have a higher percentage of tertiary sector workers than the other countries in Table 1, that is, they are positioned as the most post-industrialized group.

Roughly in all figures, white-collar jobs are plotted in the left quadrant, and blue-collar and agricultural workers are plotted in the right quadrant. The Attributes graphs for each country show that gender segregation is relatively obvious at the lower and middle education levels. At the same time, the distance between men and women at the higher education level is closer, indicating that gender segregation is smaller. This result is similar to the distribution in the U.S. when comparing Japan and the U.S. in Chap. "Multidimensional Structure of Occupational Segregation in Japan and the United States: Comparison of Trends from 1980 to 2010". In other words, in these countries, meritocracy has been adopted, and at the higher education level, individuals are evaluated based on their achievements rather than on attributes such as gender. Thus, We can note a positive relationship between industrialization and meritocracy in this group.

5 Group 2…Japan, Italy, Hungary, Poland, and Malaysia

Then, we consider Japan (Figs. 9 and 10), Italy (Figs. 11 and 12), Hungary (Figs. 13 and 14), Poland (Figs. 15 and 16), and Malaysia (Figs. 17 and 18). Based on the results in Table 1, these countries have a slightly smaller percentage of tertiary industry

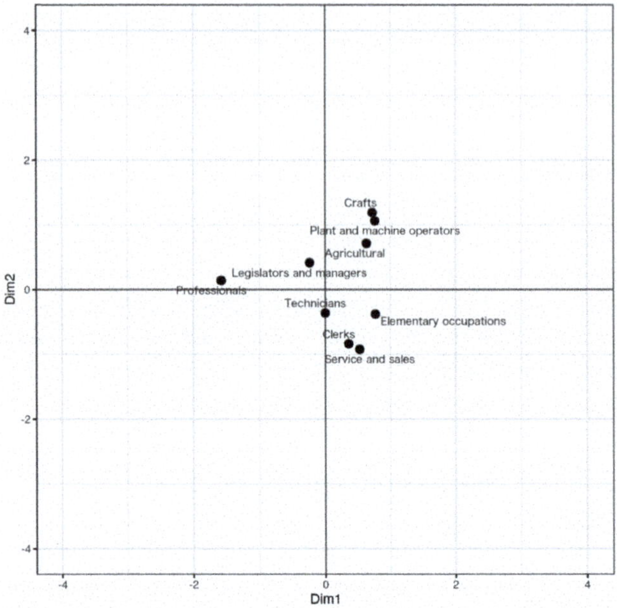

Fig. 1 The Netherlands—Occupations. *Note* The first dimension's contribution is 0.545 and the second dimension's contribution is 0.318

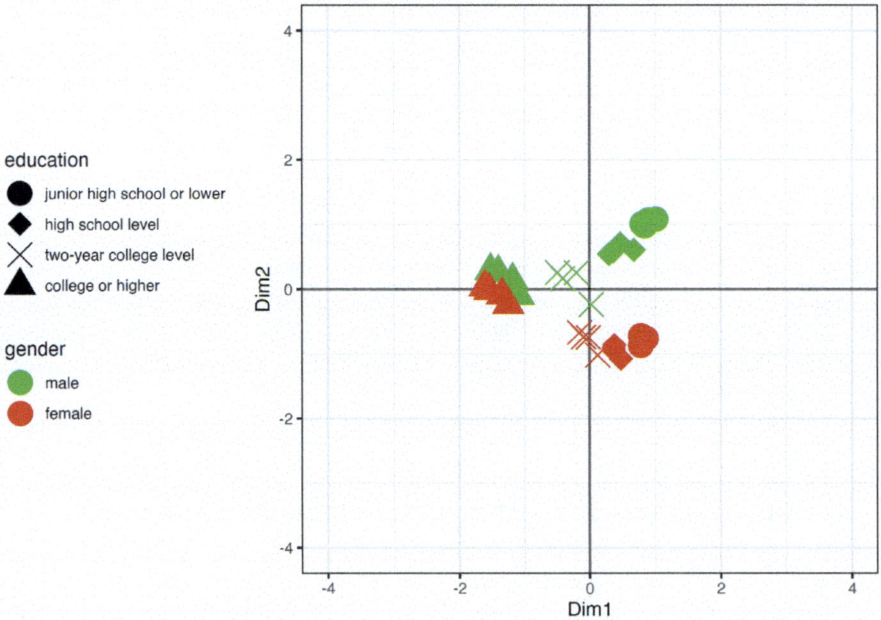

Fig. 2 The Netherlands—Attributes

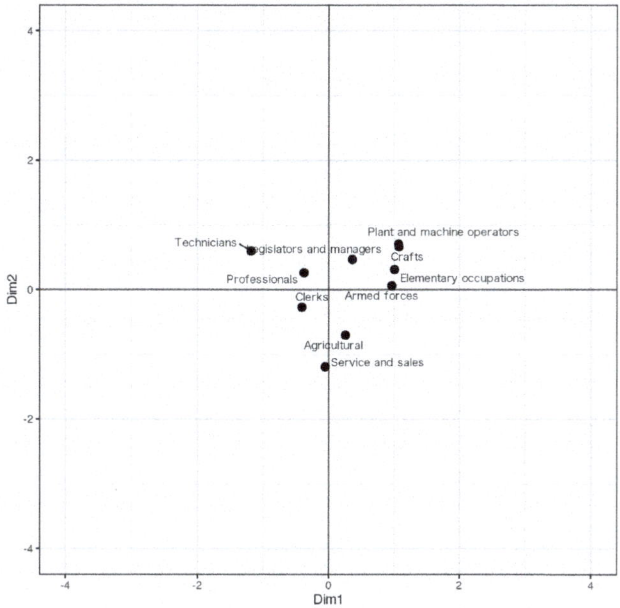

Fig. 3 The U.S.—Occupations. Note: The first dimension's contribution is 0.623 and the second dimension's contribution is 0.282

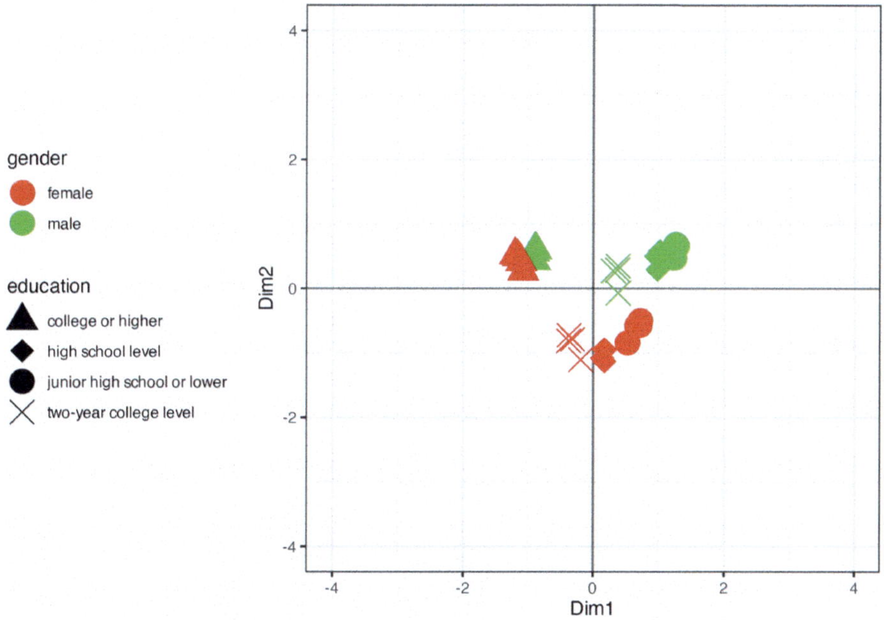

Fig. 4 The U.S.—Attributes

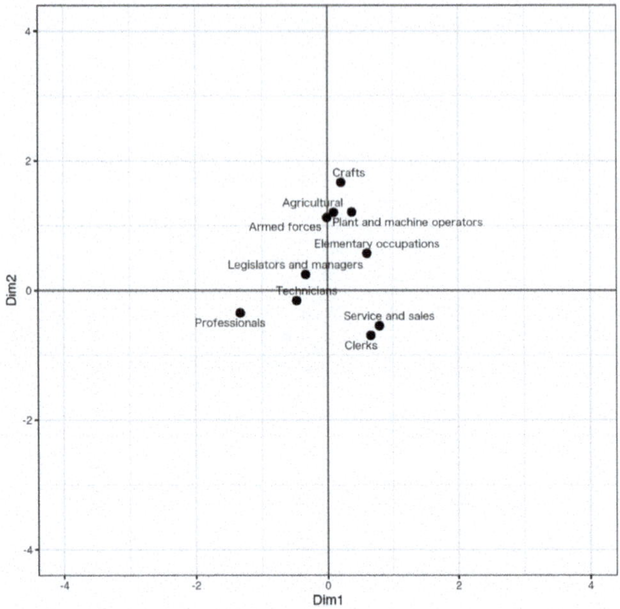

Fig. 5 The U.K.—Occupations. Note: The first dimension's contribution is 0.506 and the second dimension's contribution is 0.414

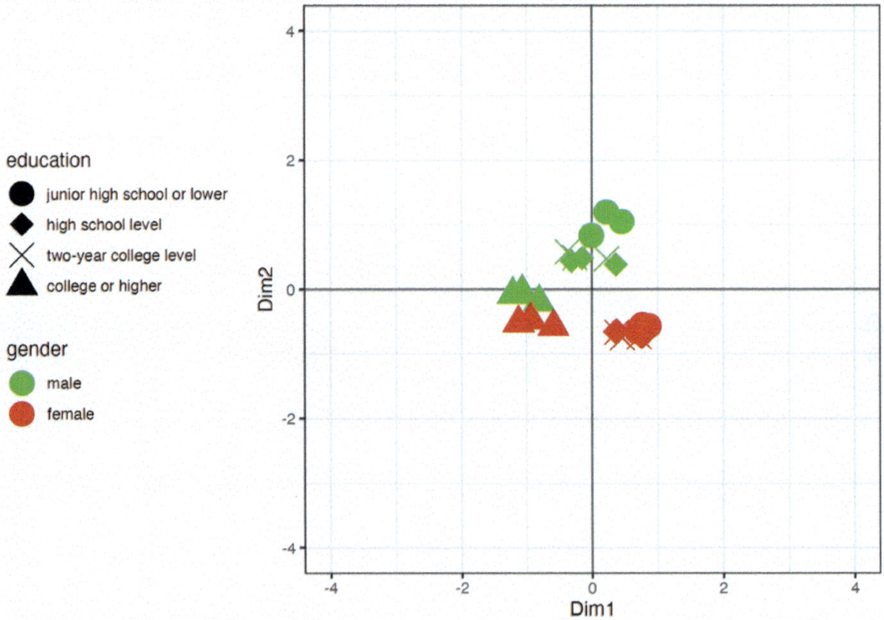

Fig. 6 The U.K.—Attributes

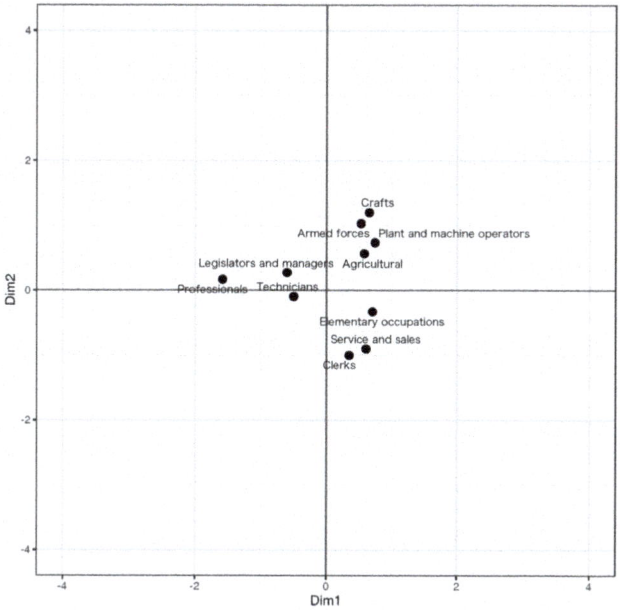

Fig. 7 France—Occupations. *Note* The first dimension's contribution is 0.507 and the second dimension's contribution is 0.366

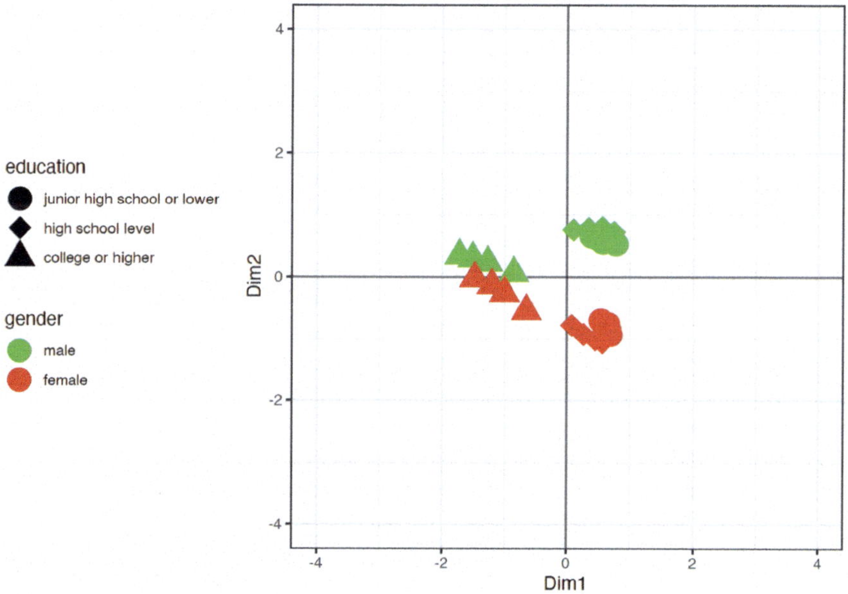

Fig. 8 France—Attributes

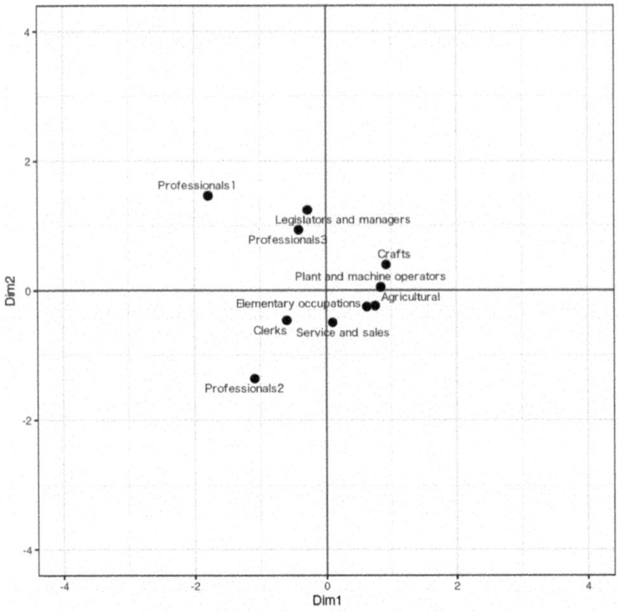

Fig. 9 Japan—Occupations. *Note* The first dimension's contribution is 0.443 and the second dimension's contribution is 0.324

than the previously mentioned groups. Thus, they can be considered a group with a relatively delayed post-industrialization.

The pattern in these countries is somewhat awkward. First, these five countries were Hungary, Italy, and Malaysia, which, like the four countries mentioned above, tended to have small occupational segregation by gender among the highly educated.

However, Hungary, Italy, and Malaysia show a different picture of middle and lower education levels. In Hungary, occupational segregation by gender can be observed at middle and lower education levels. This distribution can be interpreted as similar to the Netherlands, the U.K., the U.S., and France. However, although such segregation is observed at the middle education level in Italy and Malaysia, it is relatively small at the lower level. In other words, occupational segregation by sex in Italy and Malaysia is more pronounced in the middle of the education level.

Japan and Poland show occupational segregation by gender, even at the higher education level. Moreover, these two countries also show such segregation at the middle education level. In other words, occupational segregation by gender is observed at all levels of education. We will discuss this later.

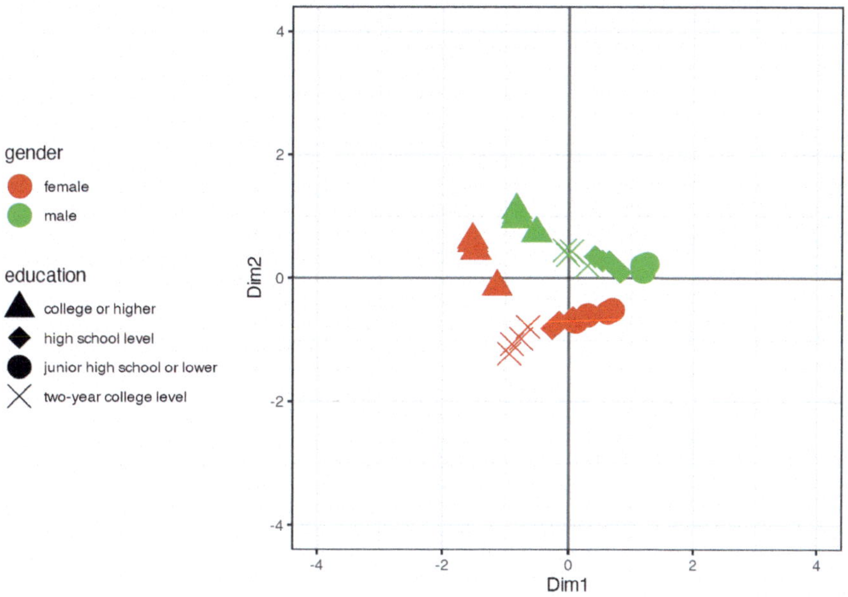

Fig. 10 Japan—attributes

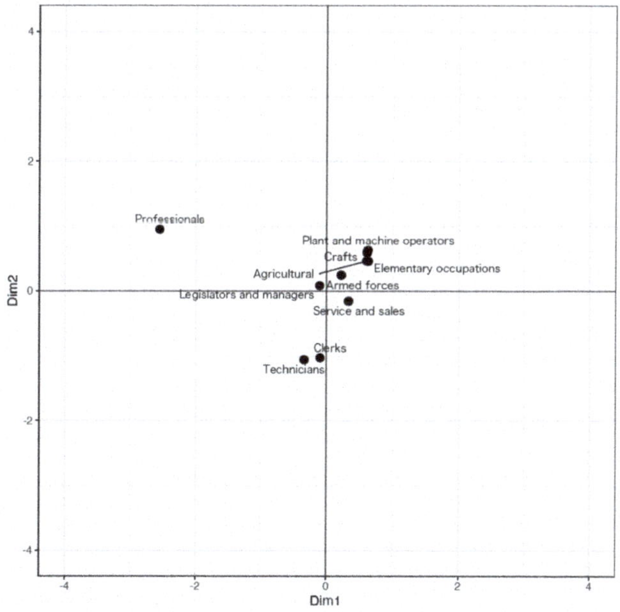

Fig. 11 Italy—occupations. *Note* The first dimension's contribution is 0.578 and the second dimension's contribution is 0.271

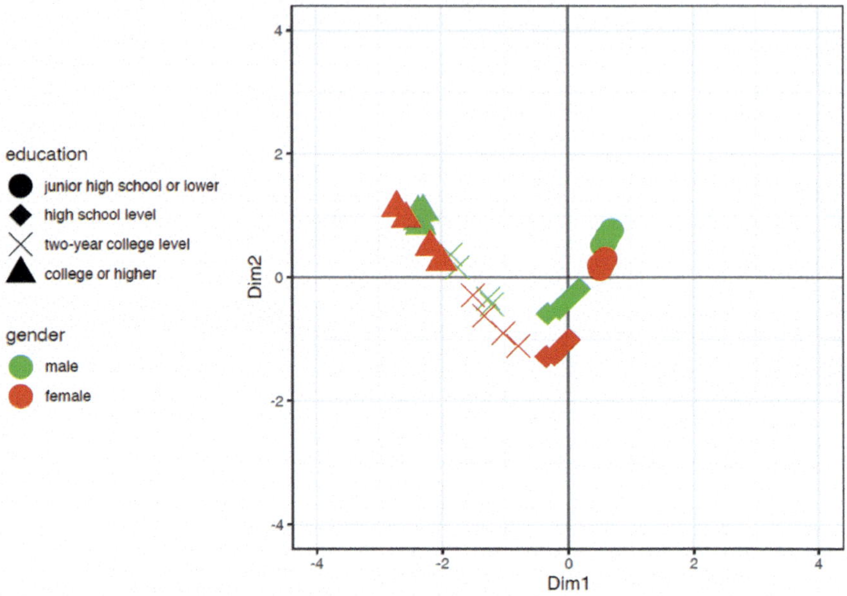

Fig. 12 Italy—attributes

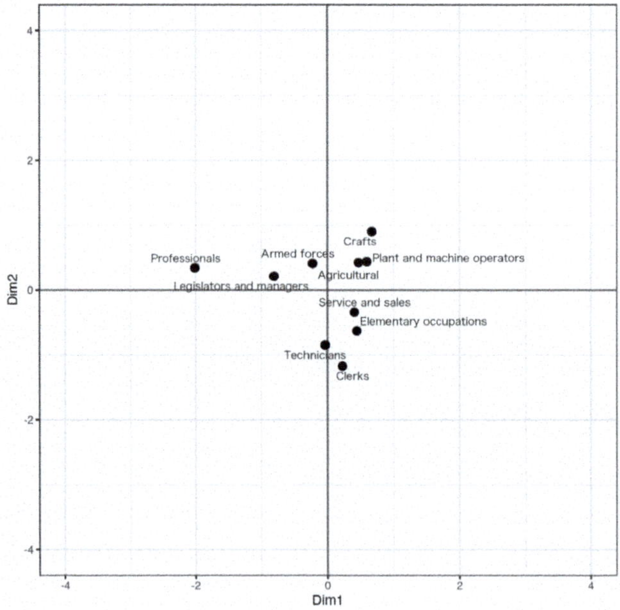

Fig. 13 Hungary—occupations. *Note* The first dimension's contribution is 0.628 and the second dimension's contribution is 0.251

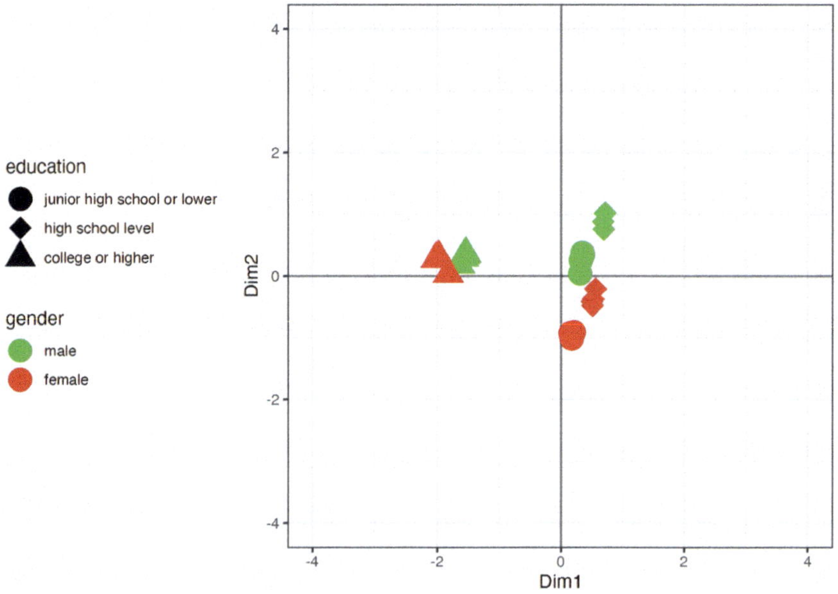

Fig. 14 Hungary—attributes

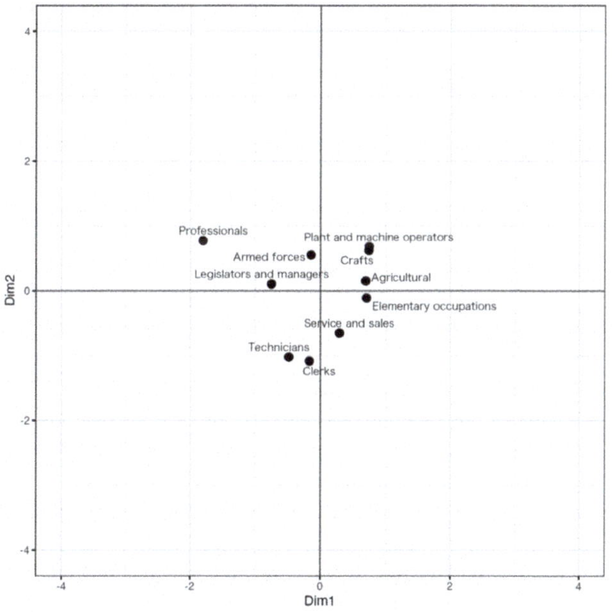

Fig. 15 Poland—occupations. *Note* The first dimension's contribution is 0.564 and the second dimension's contribution is 0.226

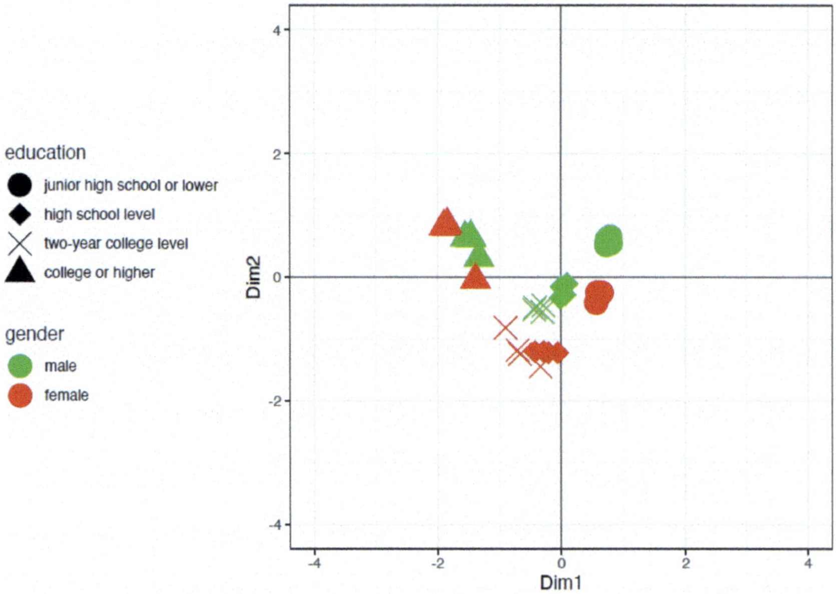

Fig. 16 Poland—attributes

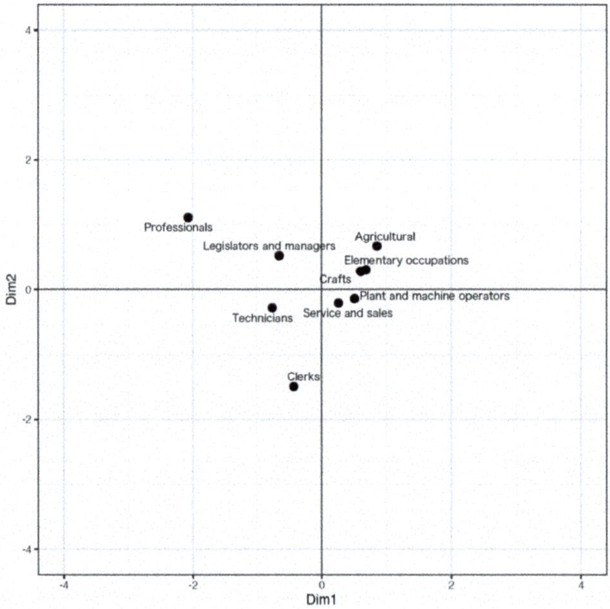

Fig. 17 Malaysia—occupations. *Note* The first dimension's contribution is 0.597 and the second dimension's contribution is 0.271

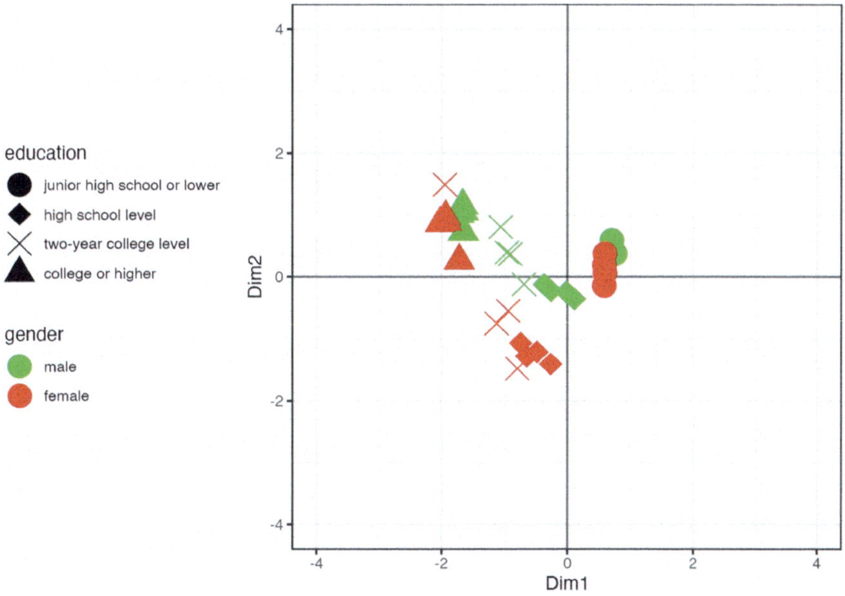

Fig. 18 Malaysia—attributes

6 Group 3…The Philippines, Thailand, China, India, and Vietnam

Next, we will examine the remaining five countries. These are the Philippines (Figs. 19 and 20), Thailand (Figs. 21 and 22), China (Figs. 23 and 24), India (Figs. 25 and 26), and Vietnam (Figs. 27 and 28). Table 1 shows that these countries have a relatively lower share of service industries and a relatively higher share of agricultural workers than the two previously mentioned groups. In other words, this group is the one that is post-industrialized later than the other two groups mentioned above.

Although they are similar in the progress of the service industry, this group is not all the same in terms of occupational segregation by gender. First, India shows relatively little occupational segregation by gender at the higher and lower education levels but prominent segregation at the middle level. In other words, the results show a similar pattern to those of Italy and Malaysia. On the other hand, Thailand, China, and Vietnam show relatively close trends. In these three countries, occupational segregation by gender is small at the lower education level, while it is more pronounced at the middle and higher education levels.

India, Thailand, China, and Vietnam all had a high percentage of agriculture. Looking at the plot of occupations corresponding to the lower education level, we can point out that agriculture is in a close position for both men and women. If so, it can be argued that occupational segregation by gender is relatively small at the lower education level in countries with a high percentage of primary industry. As

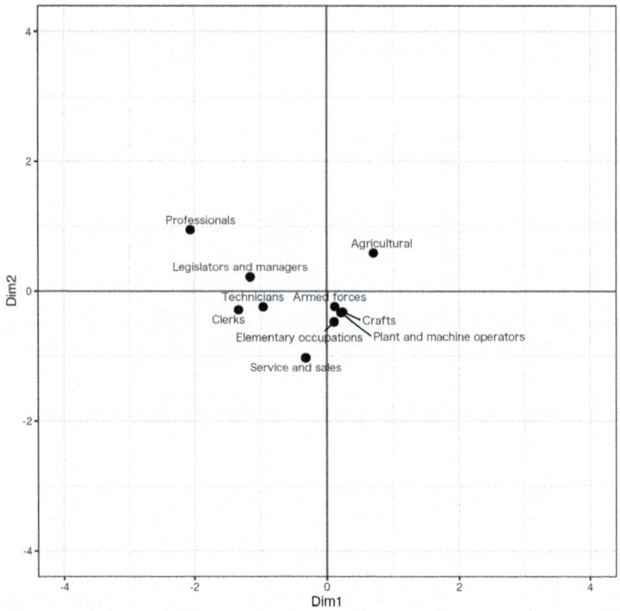

Fig. 19 The Philippines—occupations. *Note* The first dimension's contribution is 0.682 and the second dimension's contribution is 0.174

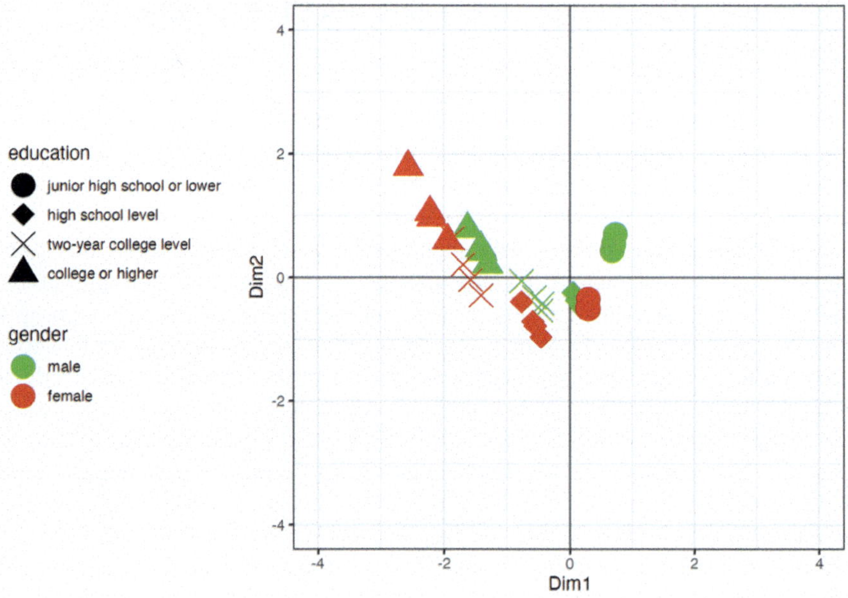

Fig. 20 The Philippines—attributes

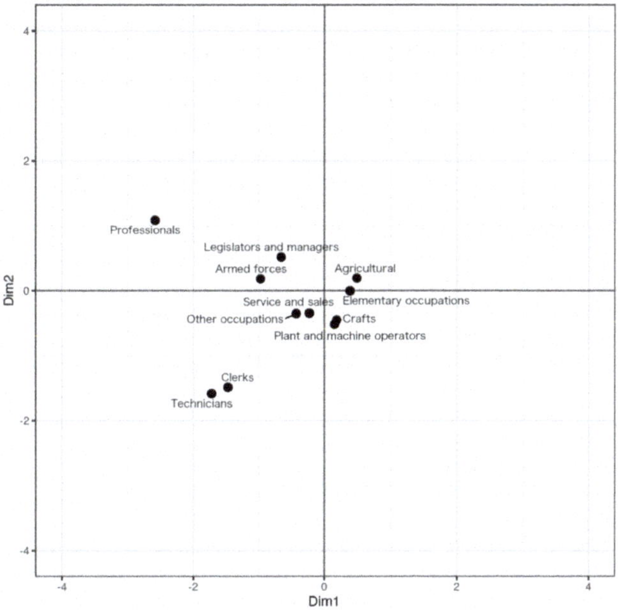

Fig. 21 Thailand—occupations. *Note* The first dimension's contribution is 0.754 and the second dimension's contribution is 0.114

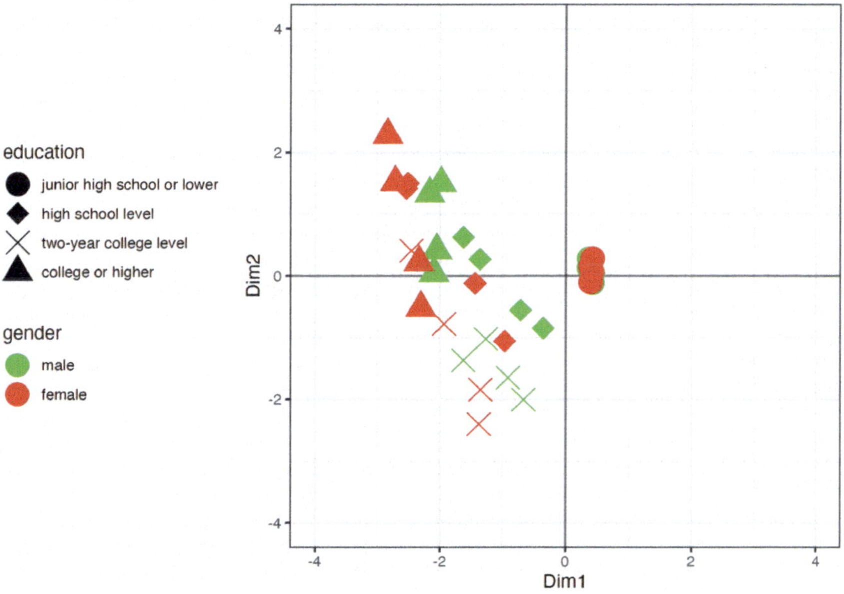

Fig. 22 Thailand—attributes

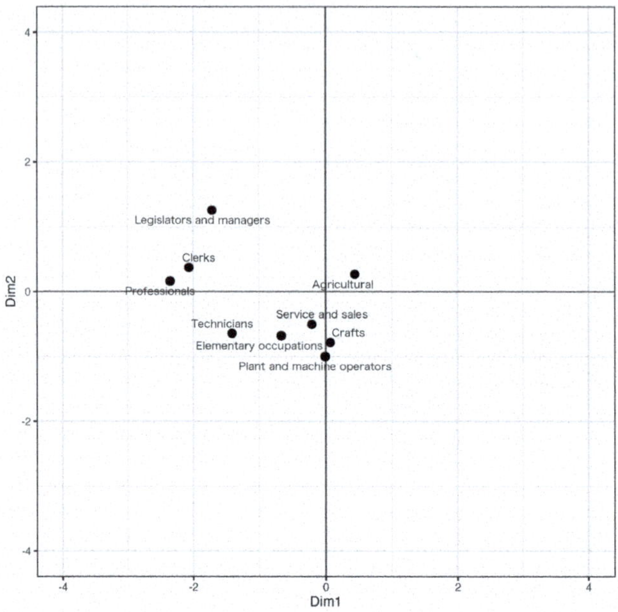

Fig. 23 China—occupations. *Note* The first dimension's contribution is 0.782 and the second dimension's contribution is 0.099

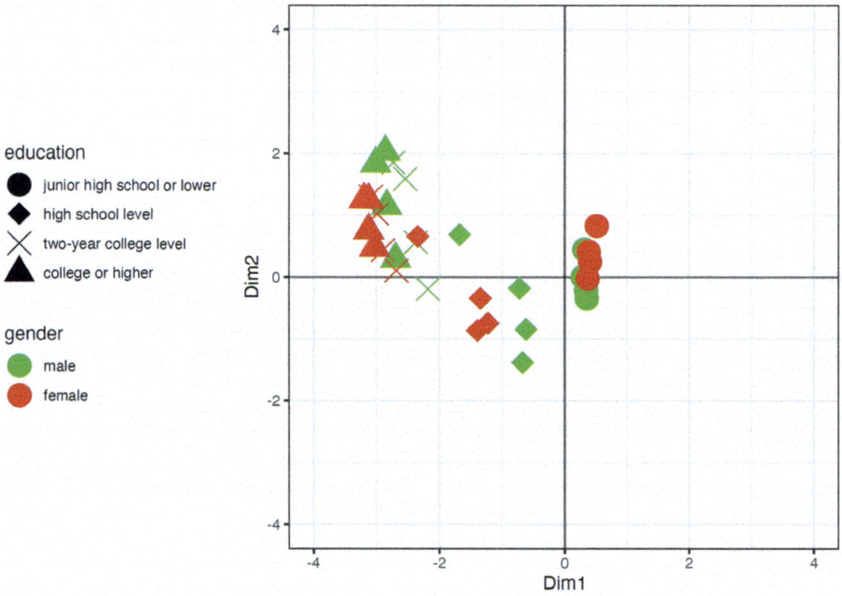

Fig. 24 China—attributes

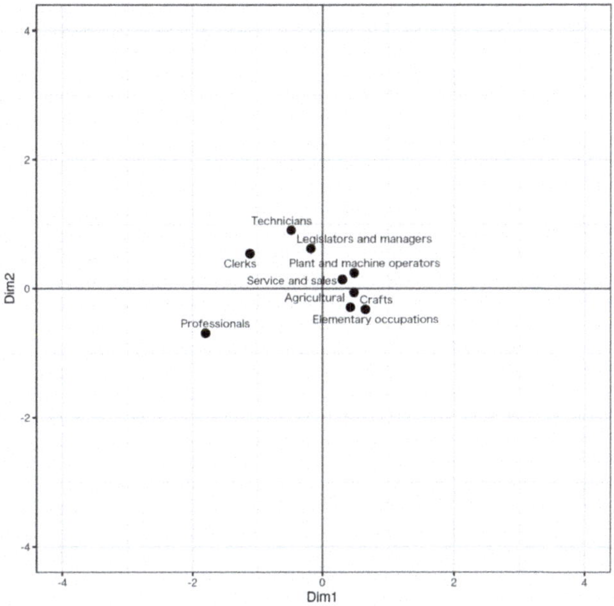

Fig. 25 India—occupations. *Note* The first dimension's contribution is 0.843 and the second dimension's contribution is 0.102

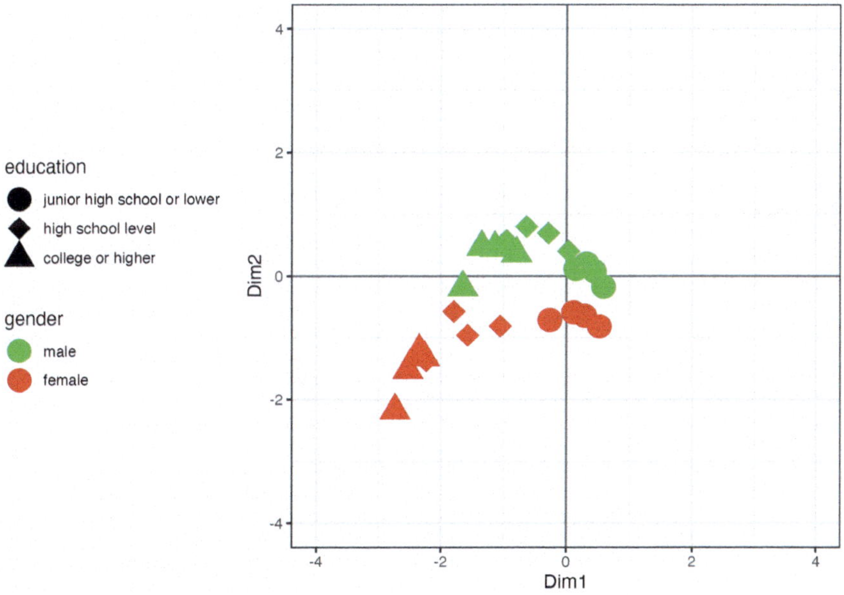

Fig. 26 India—attributes

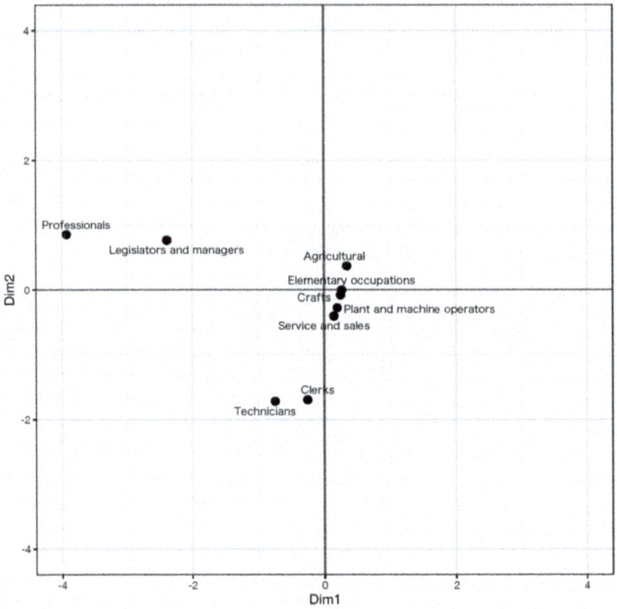

Fig. 27 Vietnam—occupations. *Note* The first dimension's contribution is 0.686 and the second dimension's contribution is 0.171

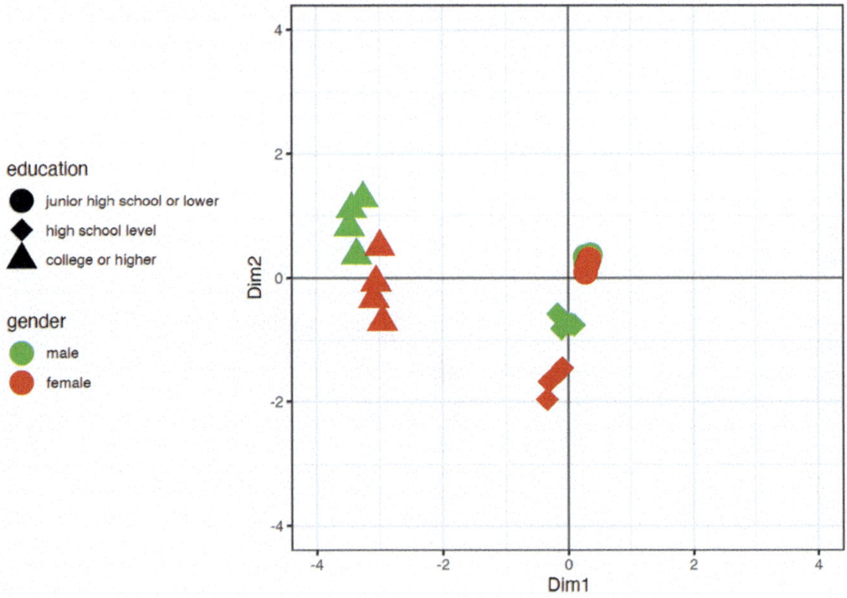

Fig. 28 Vietnam—attributes

education levels increase and occupations become more advanced and specialized, the segregation between men and women becomes more apparent. On the other hand, the Philippines shows a different pattern from those countries. In other words, occupational segregation by gender can be seen at all levels of education. Therefore, the Philippines is similar to Japan and Poland, as mentioned above.

Table 2 summarizes the 14 countries' characteristics observed in the correspondence analysis. First, the 14 countries can be roughly classified into about four groups. In Group A, occupational segregation by gender is observed at the lower and middle education levels, while segregation at the higher education level is smaller. Group A comprises the Netherlands, the U.S., the U.K., France, and Hungary. In Group B, occupational segregation is smaller at the lower and higher education levels. Instead, occupational segregation by gender is observed at the middle education level. This group comprises Italy, India, and Malaysia. Group C comprises countries with small occupational segregation by gender at the lower education level and large segregation at the middle and higher education levels. This group comprises Thailand, China, and Vietnam.

The above Group C to Group A can be considered to correspond to the progress of post-industrialized countries. In other words, Group A comprises countries where occupational segregation by gender is small in terms of higher education and where at least a meritocracy has been relatively achieved. In contrast, Group C countries can be regarded as societies that have not yet achieved a meritocracy among those with higher education. Instead, in societies with a high proportion of primary industries that do not depend on education, such as agriculture, the occupational segregation between men and women is relatively small at lower levels of education. In other words, occupational segregation by sex is changing in response to the transformation of the industrial structure toward industrialization or post-industrialized society.

Group B can be regarded as countries located in the middle of the two groups. Therefore, it is not a significant issue whether India should be placed in Group C or Group B. Overall, the correspondence analysis results from Group A to Group C indicate an overlap between the transformation of the industrial structure and the penetration of meritocracy.

However, some countries have not corresponded these industrial structural changes to the penetration of meritocracy. They are Group D. Group D, which includes Japan, is a group of countries where post-industrialization has progressed. Still, meritocracy has not become widespread, and the attributes principle, such as

Table 2 Summary of analysis_occupational segregation by gender

	Lower Edu. level	Middle Edu. level	Higher Edu. level
(1) the Netherlands, the US, the UK, France, Hungary	Large	Large	Small
(2) Italy, India, Malaysia	Small	Large	Small
(3) Thailand, China, Vietnam	Small	Large	Large
(4) Poland, the Philippines, Japan	Large	Large	Large

gender, remains clearly in place at all levels of education. In other words, in Japan and some other countries in Group D, the shift in industrial structure toward post-industrialization does not correspond to the spread of meritocracy. We will discuss this point in the next section.

7 Conclusion

This chapter explores the link between occupational segregation by gender and meritocracy through international comparisons, focusing on post-industrialization. Specifically, 14 countries are compared through correspondence analysis. We used the custom-made tabulation results of Japan's Census of Population (Ministry of Internal Affairs and Communications) for the Japan analysis. For other countries, we used IPUMS. The analysis first classified the 14 countries based on their degree of post-industrialization. Then, correspondence analysis was conducted to examine the relationship between each country's meritocracy and the occupations of men and women. The analysis classified the 14 countries into four groups. Group A includes the Netherlands, the U.S., the U.K., France, and Hungary; Group B includes Italy, India, and Malaysia; Group C includes Thailand, China, and Vietnam; and Group D includes Poland, the Philippines, and Japan.

Group A is the most post-industrialized group, with a pervasive meritocracy and decreasing gender segregation at the higher education level. Group B is somewhat post-industrialized, with smaller gender segregation at the higher and lower education levels. Group C is a group of less post-industrialized countries with notable gender segregation at the middle and higher education levels. In other words, post-industrialized development and the realization of meritocracy were compatible from Group C to Group A. As post-industrialized countries become more highly educated, occupational segregation by gender becomes less pronounced. The main finding of this chapter is the confirmation of this trend in an international comparison.

However, there was one group that did not follow this trend. These were the countries in Group D, including Japan. Group D, like Group B, had some degree of post-industrialization but occupational segregation by gender at all education levels. As discussed in the introduction chapter of this book, horizontal occupational segregation by sex is said to be more evident in societies with an expanding service sector (e.g., Charles and Grusky 2004). Based on the analysis in this chapter, which is an international comparison, we can say that Group D, including Japan, clearly shows such horizontal segregation.

Of course, some studies indicate that occupational segregation by gender exists even at high educational levels in countries such as the U.S. For example, Charles and Bradley (2009) point out that even in the U.S., highly educated women accept "gendered selves," and a gender track is formed. In other words, the reality may be that all countries show some degree of horizontal segregation of occupations based on gender. However, based on the findings of this chapter's international comparison, it can be stressed that such gender tracks are even more pronounced in other countries, such as Japan. In the U.S. and some Western European countries, segregation is smaller at higher levels of education.

The fact that occupational segregation by gender becomes less pronounced at higher levels of education in many post-industrialized countries suggests a shared expectation that education can overcome the gender barrier. However, It may be an optimistic story that modernization will free people from attributes. What kind of society would come after overcoming the barriers of attributes through educational premiums? Based on the analysis in this chapter, the results suggest a game of acquiring highly skilled human resources at the global level. Around the world, highly educated human resources will respond to advanced occupations regardless of gender (Florida 2005). If this is the case, it is not hard to imagine that highly educated personnel at the global level will be able to find more advanced occupations across countries. If higher levels of education serve as a ticket to success, then it would be desirable to attain higher levels of education. As a result, it is conceivable that admission to some world-class universities and graduate schools will become the standard for the next higher level of education. The sharing of educational premiums in a post-industrialized society may accelerate endless competition at the global level. This may lead to a strengthening of the pecking order structure, with Group A at the top.

Then, we must carefully focus on Group D, which is uniquely positioned. What does it mean for a society such as Japan, which maintains strong occupational segregation by gender despite its post-industrialized situation? It is not difficult to imagine that the more competitive and efficient system will survive as globalization progresses and competition advances. During such competition, what will Japan face where gender tracks are firmly maintained? Based on the findings of this study, we refrain from making such predictions. However, it is clear that, at the very least, an international comparison that includes Japan is essential in examining the relationship between post-industrialized countries and meritocracy.

It is easy to imagine the growing demand for certain professions at the global level, typically engineers in information and communication services. This makes it an essential strategy for countries and companies to attract talented people globally for such industries and professions. Naturally, it is desirable to have highly educated people in their home countries to fill such positions. However, filling such positions in Japan is more complicated than in other countries. Gender segregation makes it challenging to allocate highly skilled human resources functionally. Would highly skilled professionals settle in a country where the performance-based system has not yet fully penetrated? In an aging society like Japan, it is essential to secure a productive population to maintain or develop the country's strength. Familism and traditional gender roles may have contributed to Japan's development in the past. However, during the transformation of the industrial structure, such pre-modern elements are expected to become a hindrance. As meritocracy becomes accepted in many countries regarding career development, it is necessary to carefully examine the consequences of a society that maintains a strong gender track.

Finally, we would like to mention the data. Although IPUMS was used in this chapter, we used custom-made census data to analyze Japan since data on Japan is not included in IPUMS. It is desirable to use internationally formatted data such as the IPUMS for international comparisons. Although this study suggested Japanese society's lateness in terms of modernization, there also may be a problem with the attitude of Japanese academic research or government concerning the use of open data for international comparisons. In any case, we must continue to observe the trend from the perspective of international comparisons.

Notes

1. Based on Nakao et al.'s (2016) framework, Wakita (2018) compared Japan to the U.S. in a hierarchical linear model and concluded that, although the importance of education was changing, it was not weakening.
2. For a detailed explanation of PUMS and IPUMS, see "Multidimensional Structure of Occupational Segregation in Japan and the United States: Comparison of Trends from 1980 to 2010".

Acknowledgements The data for countries except Japan in this chapter were provided by IPUMS (https://www.ipums.org/). The custom-made tabulation results of the Census of Population (Ministry of Internal Affairs and Communications) in Japan were provided by the National Center for Statistics, Japan. In addition, we wish to acknowledge the statistical office that provided the underlying data in this research. We would like to thank each original member of the project, Takuya Hayashi, Aya Wakita, and Yuya Saitoh, for their excellent contribution. Yuki Kuwana joined our project later. This research received grants from JSPS: KAKENHI Grant numbers 15K03818, 18K01966, and 17K17987.

References

Blau, Peter M., and Otis Dudley Duncan, with the collaboration of Andrea Tyree. 1967. The process of stratification. In Grusky, David B., and Katherine R. Weisshaar (Eds.), *Social Stratification 4th edition*, Boulder: Westview Press, pp. 486-497.

Chang, Kyung-Sup. 1999. Compressed modernity and its discontents: South Korean society in transition. *Economy and Society* 28 (1): 30–50.

Chang, Kyung-Sup. 2010. *South Korea under compressed modernity: Familial political economy in transition*. London: Routledge.

Charles, Maria, and David B. Grusky. 2004. *Occupational Ghettos*. Stanford: Stanford University Press.

Charles, Maria, and Karen Bradley. 2009. Indulging our gendered selves? Sex segregation by field of study in 44 countries. *American Journal of Sociology* 114 (4): 924–976.

Featherman, David L., F. Lancaster Jones, and Robert M. Hauser. 1975. Assumptions of social mobility research in the U. S.: The case of occupational status. *Social Science Research* 4: 329–360.

Florida, Richard L. 2005. *Cities and the creative class*. New York: Routledge.

International Labour Organization. 2022. ILOSTAT. https://ilostat.ilo.org/data/.

IPUMS-International. 2018. Available at: https://international.ipums.org/international/. Accessed 7th January 2019.

Kuwana, Yuki. 2023. *Comparison of occupational segregation in East Asia: Analyzing the structure of segregation by educational background and gender*. Nakao. This Book Title: Springer.

Mi., Kondohelle Jackson, John H. Goldthorpe, and Colin Mills. 2005. Education, employers and class mobility. *Research in Social Stratification and Mobility* 23: 3–33.

Nakao, KeikoF, Takuya Hayashi, Aya Wakita and Yuya Saitoh. (2016) Trends in the effect of education on occupational attainment in Japan. American Sociological Association Annual Meeting, Seattle.

Saitoh, Yuya. 2023. *Multidimensional structure of occupational segregation in Japan and the United States: Comparison of trends from 1980 to 2010*. Nakao. This Book Title: Springer.

Torche, Florencia. 2011. Is a college degree still the great equalizer? Intergenerational mobility across levels in the United States. *American Journal of Sociology* 117: 763–807.

Treiman, Donald J. 1970. Industrialization and social stratification. *Sociological Inquiry* 40 (2): 207–234.

Trow, Martin. 1976. *The University in the highly educated society: From Elite to mass higher education (in Japanese)*. Tokyo: Tokyo University Press.

Wakita, Aya. 2018. Changes in educational premium: A comparison of the education-occupation linear relationship in Japan and the US. The effect of education on society reconsidered (Reports Grant number 15K03818), pp. 5–20.

Summary

Keiko Nakao and Shigemi Ohtsuki

In a contemporary world where the division of labor at the global level is advancing, many countries experience a shift in industrial structure toward the direction of industrialization and then post-industrialization. How will the macro-level changes in social structure, such as industrialization, relate to the individuals' attributes and statuses? Will this linkage between status and educational achievement be equally observed in all societies? This was primarily the focus of this study.

The features of this book can be summarized as follows. First, based on previous research emphasizing the linear relationship between occupational status and educational level, we explored the relationship by decomposing occupations into multidimensions other than status. The actual situation of meritocracy in developed societies (post-industrial society). With this new approach, we examined several countries, including Japan and the U.S.

Chapter "Multidimensional Structure of Occupational Segregation by Education and Gender: Time Series Comparison in Japan from 1980 to 2010" showed the changes in Japanese society over the past 30 years. It has been stable, based on a decomposition of occupation into multiple dimensions. At the same time, however, we noticed that the importance of education is declining. Moreover, gender is becoming increasingly important. The meritocratic view, in which education is the ticket to success, has become less persuasive in recent years. Instead, individual attributes such as gender are becoming more critical in a post-industrial society. At least, it has become clear gender segregation is stable in post-industrial societies such as Japan and has become more prominent in recent years.

In light of this point, we went on to investigate in more details the aspects of occupation and gender segregation. Chapter "Occupational Segregation by Education and Gender in Japan: Focusing on Employment Status and Marital Status"

K. Nakao (✉) · S. Ohtsuki
Tokyo Metropolitan University, 1-1 Minamiosawa, Hachioji-City, Tokyo 192-0397, Japan
e-mail: nakao@tmu.ac.jp

K. Nakao (ed.), *Multidimensional Aspects of Occupational Segregation*,
Behaviormetrics: Quantitative Approaches to Human Behavior 18,
https://doi.org/10.1007/978-981-99-8513-5_8

contains the following two findings. First, gender segregation occurs early in the career stage. Second, gender segregation occurs in an occupational category even for the same professional occupation. These characteristics of Japanese society again support the need to segment occupational categories, such as micro-class analysis. In chapters "Multidimensional Structure of Occupational Segregation by Education and Gender: Time Series Comparison in Japan from 1980 to 2010" and "Occupational Segregation by Education and Gender in Japan: Focusing on Employment Status and Marital Status", we noted that the importance of gender as an attribute is becoming increasingly apparent in a modernizing society.

Chapter "Occupation Segregation among Immigrants in Japan: Using Custom–Made Census Data in 2010" focused on nationality and ethnicity as attributes besides gender. The distortion between gender and achievement principles is further complicated by the multiple layers of immigrant groups. This may suggest that ethnicity and nationality require more attention than ever in examining the relationship between attributes and achievement in Japanese society. In so doing, paying attention to national institutions is necessary, especially as pointed out in chapter "Occupation Segregation for Immigrants in Japan: Using Custom-Made Census Data in 2010". This is because state controls, including residency status, may cancel the spread of the achievement principle.

The analysis of Japanese society in chapters "Multidimensional Structure of Occupational Segregation by Education and Gender: Time Series Comparison in Japan from 1980 to 2010" through "Occupation Segregation among Immigrants in Japan: Using Custom-Made Census Data in 2010" shows that, as Japanese society progresses toward deindustrialization, attributes rather than social stratification are becoming more pronounced. This phenomenon is paradoxical to the logic of modernization. Is this phenomenon unique to Japan? The question of this study needs to be examined from an international comparative perspective. In chapter "Multidimensional Structure of Occupational Segregation in Japan and the United States: Comparison of Trends from 1980 to 2010", we compared Japan and the United States. In the U.S., education was more clearly associated with the occupation than gender. This result contrasts with Japan, where the importance of gender attributes is becoming prominent. Chapter "Comparison of Occupational Segregation in East Asia: Analyzing the Structure of Segregation by Educational Background and Gender" is an international comparison of three East Asian countries. The analysis in this chapter shows that occupational segregation by gender is relatively more pronounced than in Japan. Japan's gender-based occupational segregation is relatively clear.

In chapter "An International Comparison of Occupational Segregation by Gender: Through a Comparison of 14 Countries", we examined the characteristics of Japanese society by making international comparisons with more nations. In this chapter, we discussed the global trend toward post-industrialization and the trend toward a meritocracy. However, some countries show unique tendencies within this trend, and Japan is a typical example. Compared with other countries, we found a rigid gender track in Japanese society, which has often been described as the "lost 30 years" after the collapse of the bubble economy. It is easy to imagine Japan will be increasingly disadvantaged in global economic competition.

What kind of policies are desired if we seek to shift society from the attributed principle to the achievement principle? Brinton analyzed gender segregation in Japan from an international comparative perspective and made the following recommendations. They are the establishment of childcare services, reform of the tax system for married couples, a social environment that makes it easier for men to take childcare leave, and the establishment of a society in which men can actively choose to work in various ways, such as shorter hours and limited working hours and work locations. As Brinton (2001) argues, the social systems designed to make men more likely to share household duties are necessary to assist women's societal advancement. In addition, this book emphasizes the need for a broader discussion of Japan's unique situation. The international comparisons show that gender segregation in Japan is not standard. Therefore, it is crucial for people living in Japanese society to be aware of the situation in Japan through cross-natioal comparisons. In this regard, immigrants may provide an opportunity for such a discussion; as indicated in chapter "Occupation Segregation among Immigrants in Japan: Using Custom-Made Census Data in 2010", meritocracy is reflected more in some immigrants of certain nationalities than in Japanese women. The fact that immigrant women, who are also highly educated, are achieving higher occupational status in Japanese society than Japanese women may prompt us to take a closer look at the gender track in Japan. In this sense, social integration in Japan may lead to a change in gender attitudes. Such changes will help Japanese society become more justified and exhaustive in an increasingly competitive global world. From now on, we must closely pay attention to the future changes in Japanese society.

Acknowledgements We would like to thank each original member of the project for their excellent contribution, Takuya Hayashi, Aya Wakita, and Yuya Saitoh. Yuki Kuwana joined our project later. This research received grants from by JSPS: KAKENHI Grant numbers 15K03818 and 18K01966 and 17K17987.

Reference

Brinton, Mary C. 2001. Married women's labor in East Asian economies. In *Women's working lives in East Asia*, ed. Mary C. Brinton, 1–37. Stanford: Stanford University Press.

The manufacturer's authorised representative in the EU is Springer Nature Customer Service Centre GmbH, Europaplatz 3, 69115 Heidelberg, Germany. If you have any concerns regarding our products, please contact ProductSafety@springernature.com

Printed and bound by CPI Group (UK) Ltd, Croydon, CR0 4YY
01/10/2026
02225945-0011